Current Topics in Microbiology

Volume 370

For further volumes:
http://www.springer.com/series/82

Jürgen A. Richt · Richard J. Webby
Editors

Swine Influenza

Responsible Series Editor: Richard W. Compans

Editors
Jürgen A. Richt
Diagnostic Medicine/Pathobiology
Center of Excellence for Emerging
and Zoonotic Animal Diseases
College of Veterinary Medicine
Kansas State University
Manhattan, KS
USA

Richard J. Webby
Division of Virology
Department of Infectious Diseases
St. Jude Children's Research Hospital
Memphis, TN
USA

ISSN 0070-217X
ISBN 978-3-642-43986-5 ISBN 978-3-642-36871-4 (eBook)
DOI 10.1007/978-3-642-36871-4
Springer Heidelberg New York Dordrecht London

Printed on acid-free paper

Springer is part of Springer Science+Business Media (www.springer.com)

Preface

From the first detailed clinical description of the disease in the Midwestern United States in 1918, to the isolation of the causative agent, the first of any influenza virus, in 1930 (Shope 1931) to its role in the genesis of the 2009 human pandemic (Garten et al. 2009), swine have played a central role in the ecology of influenza. Although not considered the major natural reservoir for influenza A viruses, that distinction belongs to aquatic waterfowl, swine are host to a limited but dynamic assortment of viruses (Webster et al. 1992). A number of subtypes of influenza A viruses of human and avian origin, including H1, H2, H3, H4, H5, H7, and H9, have been isolated from global swine populations (reviewed in Brockwell-Staats et al. 2009). Most of these isolations have, however, been limited in number and it is only H1 and H3 influenza viruses that are known to have formed stable lineages in swine. In this respect, swine influenza viruses (SIV) are similar to their counterparts in humans where H1 and H3 viruses have also been maintained. The nature of these H1 and H3 viruses differs between the two host populations, however, and, as discussed throughout this book, are even different in swine populations in different geographic regions of the world due to multiple introductions of avian and human influenza viruses.

The dynamic nature of SIV poses difficulties for the swine industry as a recurring respiratory disease, and also for public health as a source of zoonotic infection. Human infections with SIV have been recorded regularly since the introduction of more routine testing in humans. Many of these zoonotic events have occurred in instances where humans and swine are in close contact and have typically been dead-end events with little to no further spread in humans. The virologic features of SIV that limit their spread in humans are largely unknown, but the host range barrier between human and swine highlights the fact that adaptation of a virus in one mammalian host does not necessarily mean that it is well adapted to replication in another (Landolt et al. 2003). This observation is somewhat in conflict with earlier dogmas in influenza where it was suggested that mammalian passage of avian influenza viruses was a prerequisite for the emergence of human pandemics. Swine were often identified as this mammalian host due to a number of factors including the limited number of other described natural mammalian hosts and the fact that swine appeared unique in having the receptors preferred by both human and avian influenza viruses (Ito et al. 1998). The observation that swine appeared uniquely susceptible to avian and human viruses and that avian viruses grew poorly in humans led to the postulation that these animals were the mixing vessel for human pandemic viruses; and for a number of years popular thinking, without much definitive proof, was that the 1957 and 1968 human pandemics likely arose in pigs (Scholtissek et al. 1978). Subsequent human infections with H5N1, H7, and H9N2 viruses with domestic poultry as the likely source and realizations that swine were not unique in their ability to harbor avian and human viruses shifted thinking toward poultry being as important as swine as reservoirs of viruses with pandemic potential. Indeed, the global spread of highly pathogenic H5N1viruses focused a lot of research effort and funding toward avian hosts at the expense of solidifying activities in swine. Although surveillance and research activities of influenza in swine continued, and to some degree increased, during the first decade of the twenty-first century, these activities were dwarfed by the efforts going on in wild and domestic poultry species. The isolation of a novel influenza virus (i.e., pandemic H1N1) from a 10-year-old boy in California in April 2009 indicated that more of the influx of resources should have been funneled into further understanding the global SIV situation. The virus from the 10-year-old was obviously of SIV ancestry, but it was different enough from any other virus characterized that its direct precursors still remain a mystery. In addition, in 2012 zoonotic transmission of SIV (both H3N2 and H1N2 subtypes) containing the matrix gene from the pandemic H1N1 virus was reported. These strains appeared to be able to spread more easily from pigs to people than other influenza viruses of swine. More than 300 people from 10 states were reported to have been infected with these new strains resulting in hospitalizations and 1 death; limited human-to-human transmission was detected (Lindstrom et al. 2012). Importantly, the main risk factor for infection was exposure to pigs, mostly in the context of agricultural fair settings.

With these events firmly at center stage, it is a good opportunity to review what we know about SIV as a disease of swine and also as a continued zoonotic threat. The 15 chapters presented in this book provide contemporary reviews of research on SIV. The book begins with a general overview of influenza viruses by Stephan Pleschka discussing the virus and its replication in detail. The history of SIV in North America, Europe, and Asia is discussed by Stacey Schultz-Cherry, Christopher Olsen, and Bernard Easterday, by Roland Zell, Christoph Scholtissek, and Stephan Ludwig, and by Huachen Zhu, Richard Webby, Tommy Lam, David Smith, Malik Peiris, and Yi Guan, respectively. As indicated in these reviews, the European, North American, and Asian SIV evolution follows different pathways. Whereas descendants of classical SIV and the novel triple reassortant viruses are found in North America, avian-like swine H1N1 viruses emerged in Europe in 1979 after an avian to swine transmission and spread to all major European pig-producing countries where they circulate with H3N2 and H1N2 reassortants. Classical swine H1N1, human-origin H3N2, avian-like H1N1 and the triple reassortant viruses all co-circulate in Asian pigs. The clinicopathological features of SIV infections in pigs are described by Bruce Janke. Macroscopic and microscopic lesions of SIV infection, after natural and experimental infection, are described. The use of accurate diagnostics assays for diagnosis and surveillance for SIV are summarized by Susan Detmer, Marie Gramer, Sagar Goyal, Montserrat Torremorell, and Jerry Torrison. Since our collective knowledge regarding the worldwide occurrence of influenza among swine is incomplete, this review focuses on basic laboratory assays needed for the detection of the virus and viral nucleic acids within clinical samples and for antiviral antibodies in serum samples.

The epidemiology of swine influenza worldwide is of exceptional importance with the potential of the pig acting as a "mixing vessel" where both avian and human influenza viruses can undergo genetic reassortment resulting in the creation of novel viruses. The reviews by Alessio Lorusso, Amy Vincent, Marie Gramer, Kelly Lager, and Janice Ciacci-Zanella on North American, by Ian Brown on European, and by Young-Ki Choi, Philippe Noriel Pascua, and Min-Suk Song on Asian swine influenza epidemiology shed light on how this unique ability of pigs results in ever expanding new genotypes and subtypes in pigs. Vaccination is still one of the most important and effective strategies to prevent and control influenza for both the animal and human population. The review by Kristien van Reeth and Wenjun Ma discusses the current and future options to control this economically important swine disease.

The zoonotic aspects of SIV infections are reviewed by Whitney Baker and Gregory Gray. Most of these infections have been sporadic cases with a recent increase of case reports in concert with modern pig farming and the emergence of triple reassortant SIV. The advent of pandemic H1N1 and its impact on human health is discussed by Ian York and Ruben Donis, while Julia Keenliside discusses its impact on animal populations. Hadi Yassine, Chang-Won Lee, and Yehia Saif describe another important interspecies transmission event of influenza A viruses, namely the one between swine and poultry. Swine viruses are continuously isolated from poultry species, especially turkeys, and they are causing economic losses. Finally, Elena Govorkova and Jonathan McCullers cover the critical area of approved and investigational antiviral drugs.

We would like to thank the contributors for their patience during the assembly of this volume. We hope that all readers will gain insight from these contributions that will enhance their individual research and teaching activities.

J. A. Richt
R. J. Webby

References

Brockwell-Staats C, Webster RG, Webby RJ (2009) Diversity of influenza viruses in swine and the emergence of a novel human pandemic influenza A (H1N1). Influenza Other Respi Viruses 3:207–213

Garten RJ, Davis CT, Russell CA, Shu B, Lindstrom S, Balish A, Sessions WM, Xu X, Skepner E, Deyde V, Okomo-Adhiambo M, Gubareva L, Barnes J, Smith CB, Emery SL, Hillman MJ, Rivailler P, Smagala J, de Graaf M, Burke DF, Fouchier RA, Pappas C, Alpuche-Aranda CM,

López-Gatell H, Olivera H, López I, Myers CA, Faix D, Blair PJ, Yu C, Keene KM, Dotson PD Jr, Boxrud D, Sambol AR, Abid SH, St George K, Bannerman T, Moore AL, Stringer DJ, Blevins P, Demmler-Harrison GJ, Ginsberg M, Kriner P, Waterman S, Smole S, Guevara HF, Belongia EA, Clark PA, Beatrice ST, Donis R, Katz J, Finelli L, Bridges CB, Shaw M, Jernigan DB, Uyeki TM, Smith DJ, Klimov AI, Cox NJ (2009) Antigenic and genetic characteristics of swine-origin 2009 A(H1N1) influenza viruses circulating in humans. Science 325:197–201

Ito T, Couceiro JN, Kelm S, Baum LG, Krauss S, Castrucci MR, Donatelli I, Kida H, Paulson JC, Webster RG, Kawaoka Y (1998) Molecular basis for the generation in pigs of influenza A viruses with pandemic potential. J Virol 72:7367–7373

Landolt GA, Karasin AI, Phillips L, Olsen CW (2003) Comparison of the pathogenesis of two genetically different H3N2 influenza A viruses in pigs. J Clin Microbiol 41:1936–1941

Lindstrom S, Garten R, Balish A, Shu B, Emery S, Berman L, Barnes N, Sleeman K, Gubareva L, Villanueva J, Klimov A (2012) Human infections with novel reassortant influenza A(H3N2)v viruses, United States, 2011. Emerg Infect Dis 18:834–837

Scholtissek C, Rohde W, Von Hoyningen V, Rott R (1978) On the origin of the human influenza virus subtypes H2N2 and H3N2. Virology 87:13–20

Shope RE (1931) Swine Influenza. III. Filtration experiments and aetiology. J Exp Med 54:373–385

Webster RG, Bean WJ, Gorman OT, Chambers TM, Kawaoka Y (1992) Evolution and ecology of influenza A viruses. Microbiol Rev 56:152–179

Contents

Overview of Influenza Viruses

Stephan Pleschka

Abstract The influenza virus (IV) is still of great importance as it poses an immanent threat to humans and animals. Among the three IV-types (A, B, and C) influenza A viruses are clinically the most important being responsible for severe epidemics in humans and domestic animals. Aerosol droplets transmit the virus that causes a respiratory disease in humans that can lead to severe pneumonia and ultimately death. The high mutation rate combined with the high replication rate allows the virus to rapidly adapt to changes in the environment. Thereby, IV escape the existing immunity and become resistant to drugs targeting the virus. This causes annual epidemics and demands for new compositions of the yearly vaccines. Furthermore, due to the nature of their segmented genome, IV can recombine segments. This can eventually lead to the generation of a virus with the ability to replicate in humans and with novel antigenic properties that can be the cause of a pandemic outbreak. For its propagation the virus binds to the target cells and enters the cell to replicate its genome. Newly produced viral proteins and genomes are packaged at the cell membrane where progeny virions are released. As all viruses IV depends on cellular functions and factors for their own propagation, and therefore intensively interact with the cells. This dependency opens new possibilities for anti-viral strategies.

S. Pleschka (✉)
Institut für Medizinische Virologie, Justus-Liebig-Universität Gießen,
Schubertstr. 81, 35392 Gießen, Germany
e-mail: stephan.pleschka@viro.med.uni-giessen.de

Current Topics in Microbiology and Immunology (2013) 370: 1–20
DOI: 10.1007/82_2012_272

Published Online: 3 November 2012

Contents

1 Introduction

Influenza viruses (IVs) are a continuous and severe global threat to mankind and many animal species. The resulting disease gives rise to thousands of deaths and enormous economic losses in livestock each year. Clearly, influenza is a highly contagious, acute respiratory disease with global significance that affects all age groups and can occur repeatedly. Since waterfowl represents the natural reservoir for the etiological agent of the disease—the influenza and many other animal species can be infected, the virus cannot be eradicated. Therefore, a constant re-emergence of the disease will continue to occur (Palese and Shaw 2007; Webster 1999; Wilschut 2005; Wright et al. 2007). Epidemics appear in the human population almost annually and are due to an antigenic change of the viral surface glycoproteins (Fig. 1). Furthermore, highly pathogenic strains of influenza A virus have emerged unpredictably but repeatedly in recent history as pandemics like the "Spanish-Flu" that caused the death of 20–40 million people worldwide (Taubenberger et al. 2000; Webster 1999). The 2009 pandemic outbreak of the swine-origin IV (S-OIV, "Mexico-Flu") and its rapid spread around the world, as well as repeated human infections with highly pathogenic avian IV (HPAIV) of the H5-subtype demonstrated the imminent danger that IV continues to pose to both the human population and economically relevant animals.

2 The Virus and Its Replication

2.1 The Virion

IVs belong to the family of the *Orthomyxoviridae* and possess a segmented, single-stranded RNA-genome with negative orientation. IVs are divided into three types, A, B, and C based on the genetic and antigenic differences. They infect mammals and birds. Among the three types, influenza A viruses are clinically the most important pathogens and have been responsible for severe epidemics in humans and domestic animals in the past. Thus the focus of this chapter will be on type-A influenza viruses. A detailed description of the viral proteins and the replication

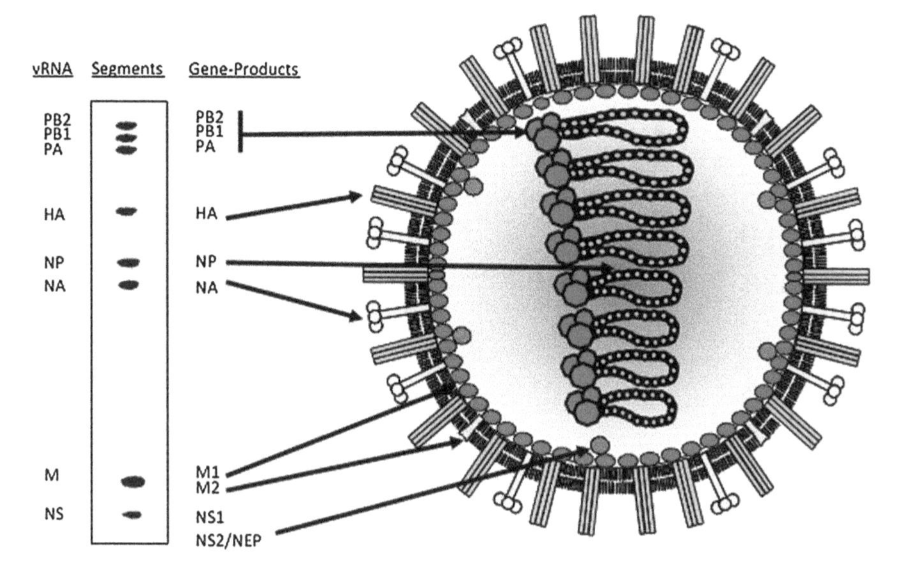

Fig. 1 The influenza A virus particle. Schematic representation of the spherical influenza A virus particle that has a diameter of about 100 nm. The eight viral RNA segments were separated by urea-polyacrylamide gel electrophoresis and visualized by silver staining (*left*). The corresponding gene products and their presumed location in the virus particle are indicated (*right*). PB1-F2 and NS1 are not a structural part of the mature virion. For details see text

cycle of influenza A viruses can be found elsewhere (Ludwig et al. 1999; Palese and Shaw 2007; Wright et al. 2007). Therefore, only an overview on these topics is given without referring to individual references.

The influenza A virus particle is composed of a lipid envelope derived from the host cell and of 9 or 10 structural virus proteins (Fig. 1 and Table 1). The components of the RNA-dependent RNA-polymerase complex (RdRp), PB2, PB1, and PA are associated with the ribonucleoprotein complex (RNP) and are encoded by the vRNA segments 1–3. The PB1 segment of many, but not all, influenza A virus strains also contains a +1-reading frame encoding the recently discovered PB1-F2 protein (Chen et al. 2001).

The viral surface glycoproteins hemagglutinin (HA) and neuraminidase (NA) are expressed from vRNA segments 4 and 6, respectively. The nucleoprotein (NP), the major component of the RNPs, is encoded by segment 5 and associates with the vRNA segments. Each of the two smallest vRNA segments code for two proteins. The matrix protein (M1) is co-linear translated from the mRNA of segment 7 and forms an inner layer within the virion. A spliced version of the mRNA gives rise to a third viral transmembrane component, the M2 protein, which functions as a pH-dependent ion channel. Employing a similar coding strategy, segment 8 harbors the sequence information for the nonstructural NS1 protein and the nuclear export protein (NEP). NEP is a minor component of the virion and is found associated with the M1 protein.

Table 1 Influenza A Virus Genome (strain A/PR/8/34)

Segment	vRNA	Protein	AA	Function(s)
1	2,341	PB2	759	Cap-binding subunit of the viral RdRp; cap-binding
2	2,341	PB1	757	Central location of the polymerase domain of the viral RdRp
		PB1-F2	87–91	Pro-apoptotic activity
3	2,233	PA	716	Cap-snatching endonuclease subunit of the viral RdRp
4	1,778	HA	566	Surface glycoprotein; receptor binding, membrane fusion
5	1,565	NP	498	Nucleoprotein; encapsidation of viral genomic and anti-genomic RNA
6	1,413	NA	454	Surface glycoprotein; receptor destroying Neuraminidase activity
7	1,027	M1	252	Matrixprotein
		M2	97	Ion channel activity, protecting HA conformation
8	890	NS1	230	Regulation of viral RdRp activity Interferon antagonist; Enhancer of viral mRNA translation; inhibition of (i) pre-mRNA splicing, (ii) cellular mRNA-polyadenylation, (iii) PKR activity,
		NEP	121	Nuclear export factor

Table 1 summarizes details of the genome segments, the encoded viral proteins and their respective function.

2.2 *The Influenza Virus Replication Cycle: Viral Proteins and Their Function*

2.2.1 Adsorption and Entry

The viral replication cycle is initiated by the binding of the HA to sialic-acid (neuraminic acid) containing cellular membrane resident molecules that act as receptors determinants. For example, it was shown that the epidermal growth factor receptor (EGFR) promotes uptake of IV into host cells (Eierhoff et al. 2010). Subsequently, the virus particle is taken up via endocytosis (Fig. 2) [For references: (Palese and Shaw 2007; Wright et al. 2007)]. Due to the different preferences for specific receptor determinants on the target cells of birds and humans, HA is regarded as a possible restriction factor. HAs of avian viruses bind to Sia2-3Gal-terminated sialylglycoconjugates, whereas those of human IV display a Sia2-6Gal-containing receptor-binding specificity [Reviewed in (Paulson 1985) see also (Connor et al. 1994)]. Nevertheless, it was recently shown that a strictly avian H7-type HPAIV carrying the NS segment of a H5-type HPAIV could

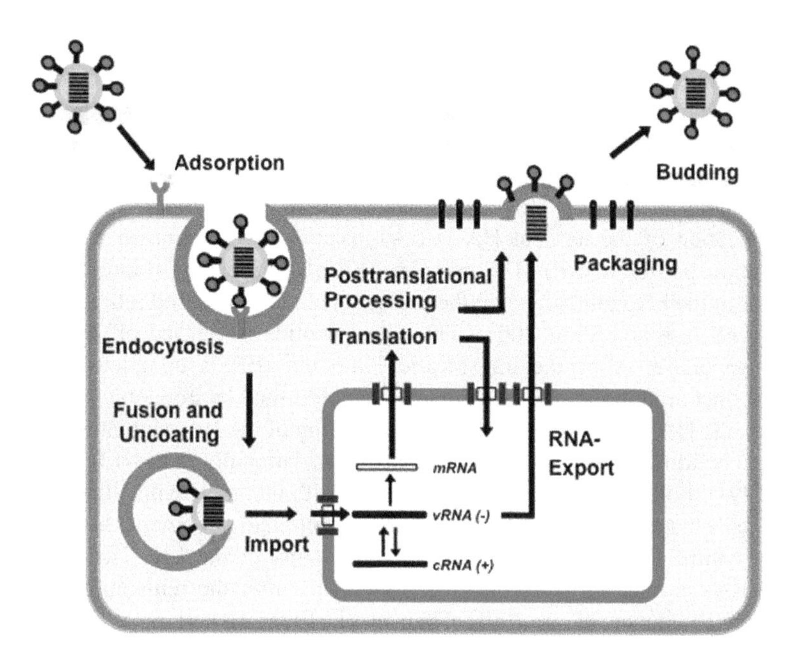

Fig. 2 The influenza viruses replication cycle. The virion attaches to the cellular receptor determinant. The receptor-bound particle enters the cell via endocytosis. After fusion of the viral and the endosomal membrane the viral genome is released into the cytoplasm. The RNPs are transported into the nucleus where replication and transcription of the viral RNA segments occur. The mRNAs are exported into the cytoplasm and are translated into viral proteins. The viral glycoproteins enter the exocytotic transport pathway to the cell surface. Replicative viral proteins enter the nucleus to amplify the viral genome. In the late stage of the infection cycle newly synthesized RNPs are exported from the nucleus and are assembled into progeny virions that bud from the cell surface

acquire the ability to replicate more efficiently in mammalian cell culture, and in contrast to the wild type was able to infect mice causing disease and death (Ma et al. 2010). Furthermore, additional NS reassortants displayed altered propagation ability of the H7-type HPAIV (Wang et al. 2010). Taken together, these results shed further light on the importance of the NS segment for viral replication, molecular pathogenicity and host range, as well as the possible consequences of a reassortment between naturally occurring H7 and H5 type HPAIVs. This indicates that the receptor HA-specificity, although important is not the sole host range and tropism determining factor.

The HA has to undergo a multitude of maturation steps, which are completely dependent on interactions with the protein processing machinery of the infected cell. To gain insight into intra-cellular post-translational processing and transport of glycoproteins, the HA has long been used as a model protein; and HA is probably the best analyzed virus component. A great amount of data has accumulated on the maturation and function of the HA during the viral replication cycle [For overviews: (Ludwig et al. 1999; Palese and Shaw 2007; Wright et al. 2007)].

The HA is a type I glycoprotein. The precursor HA_0 of the mature HA follows the exocytotic cellular transport pathway from the rER via the Golgi complex and the trans-Golgi network (TGN). In polarized epithelial cells, which represent the major viral target cell type in the respiratory and intestinal tracts, HA is transported to the apical surface and thereby defines the site of virus release (Gottlieb et al. 1986; Rindler et al. 1984; Rodriguez-Boulan et al. 1983, 1984). The *N*-terminal signal peptide of the nascent HA is co-translationally recognized by the signal recognition particle (SRP). The complex (SRP/HA/ribosome) binds to the SRP receptor in the ER membrane and the protein chain is transferred into the lumen of the rER (Palese and Shaw 2007). The signal peptide is cleaved off by a luminal signal peptidase. After the translocation into the rER is completed, the HA_0 remains anchored in the rER membrane by a C-terminal hydrophobic sequence. In the rER, the HA_0 becomes *N*-glycosylated. Folding of the HA_0 and intra-molecular disulfide bond formation occurs co- and post-translationally in the rER (Braakman et al. 1991). Folding intermediates of the HA_0 with incomplete disulfide bonds are bound by chaperones. These promote proper folding, oligomeric assembly and quality control of newly synthesized glycoproteins in the ER. The completely folded HA_0 is released from the chaperones only after the remaining glucose is removed (Braakman et al. 1991; Chen et al. 1995; Hebert et al. 1995, 1997; Peterson et al. 1995; Tatu and Helenius 1997). Misfolded HA_0 is degraded in the rER (Copeland et al. 1986; Gething et al. 1986; Hurtley et al. 1989). Properly folded HA_0 monomers assemble into trimers in the rER and are selectively transported to the *cis*-golgi compartment (Ceriotti and Colman 1990).

Another post-translational modification of the HA is acylation of conserved cysteine residues in the cytoplasmatic tail with long-chain fatty acids (Schmidt 1982; Veit et al. 1991). The corresponding acyltransferase is located in the rER (Chen et al. 2001) and the HA_0 is acylated before it reaches the Golgi. The results of several studies suggest that acylation can modulate the fusion activity of the HA (Fischer et al. 1998; Lambrecht and Schmidt 1986; Melikyan et al. 1997; Naeve and Williams 1990; Philipp et al. 1995; Simpson and Lamb 1992; Steinhauer et al. 1991), but has no major effect on the post-translational processing, intra-cellular transport, or receptor binding. There are divergent ideas about the importance of the cytoplasmic tail and its acylation for virus maturation. Reverse genetic analyses of the HA (H7 subtype) suggested that the integrity of the cytoplasmic tail and its acylation is advantageous. It was shown that acylation-mediated membrane anchoring of HA is essential for fusion pore formation and virus infectivity (Wagner et al. 2005).

Upon transport of the HA through the Golgi apparatus, the *N*-linked oligosaccharides are enzymatically processed to more complex forms by a number of different transferases. Many studies on the relevance of glycosylation for HA function indicate that glycosylation plays an important role in the virulence of influenza A viruses. Glycosylation patterns of the HA are host cell-specific and affect folding, transport, proteolytic cleavage, receptor binding, and fusion activity of the HA (Gallagher et al. 1988; Gambaryan et al. 1998; Kawaoka et al. 1984; Mir-Shekari et al. 1997; Ohuchi et al. 1997a, b; Schulze 1997) and thereby the infectivity of the virus. In addition, carbohydrate side-chains attached to the HA have been found to affect antigenic properties and modulate HA recognition by CD4 + T cells (Drummer et al. 1993; Munk et al. 1992).

The final maturation event of the HA, which renders the virus fully infectious and determines its ability to spread in the tissue of the infected host, is the proteolytic cleavage of the HA_0 into its subunits HA_1 and HA_2. This cleavage is absolutely essential for HA-function and cell infection. This cleavage primes the HA molecule to undergo a drastic conformational change in a low pH-environment. This structural rearrangement of the HA results in exposure of the hydrophobic *N*-terminus of the HA_2 peptide that is able to induce fusion between the viral and cellular membranes. The cleavage activation of HA is mediated by cellular or extra-cellular enzymes and determines both tropism and the clinical outcome of an influenza infection (Klenk and Garten 1994; Rott et al. 1995). A number of proteases have been identified that can activate the HA molecule depending on the amino acid sequence, accessibility and structure of the cleavage site. Some subtype H5 and H7 HA proteins containing several basic amino acid residues at their cleavage sites are activated by ubiquitous intra-cellular subtilisin-like eukaryotic endoproteases such as furin and PC6 (Horimoto et al. 1994; Stieneke-Grober et al. 1992; Walker et al. 1994). Due to the ubiquitous presence of these proteases, avian viruses possessing HAs with multibasic cleavage sites, can be produced in infectious form in most host organs and are therefore highly pathogenic (highly pathogenic avian IV = HPAIV). In contrast, HAs of the other subtypes always contain monobasic cleavage sites and are activated only after virus release by extracellular proteases. This includes trypsin (Klenk et al. 1975; Lazarowitz and Choppin 1975), a chicken endoprotease that shows homology to the blood clotting factor X (Gotoh et al. 1990), inflammatory active proteases such as kallikrein, urokinase, thrombin (Scheiblauer et al. 1992), and tryptase Clara which is secreted from cells of the bronchiolar epithelia (Kido et al. 1992, 1993; Sakai et al. 1994; Tashiro et al. 1992). Very often a combined infection of IV and bacteria like *Staphylococcus aureus*, *-pneumoniae* and *Hemophilus influenzae* occurs. The Staphylococcus protease is another serine protease-like enzyme that has the capacity to cleave HA and thereby accelerates the spread of a co-infecting influenza virus (Tashiro et al. 1987, 1992). Recently, novel type II transmembrane serine proteases (MSPL, TMPRSS13, and HAT) were identified that proteolytically activate HA membrane fusion activity and induce multi-cycle replication (Okumura et al. 2010; Bottcher et al. 2006).

After adsorption and endosomal uptake, virus disassembly occurs in the acidic environment of late endosomal vesicles and involves two crucial events. First, the conformation of the HA is changed to a low-pH form, which results in exposure of a fusion active protein sequence within the HA_2 to initiate fusion with the viral envelope. Next, the low pH in the endosomes activates the viral M2 ion channel protein resulting in a flow of protons into the interior of the virion. Acidification within the viral particle facilitates dissociation of the RNPs from the M1 protein. Vacuolar (H+)-ATPases (V-ATPases) facilitate these steps by acidifying the endosomal interior. The V-ATPase activity is mediated by IV-induced extracellular signal-regulated kinase (ERK) and phosphatidylinositol 3-kinase (PI3K) activity (Marjuki 2010). The RNPs are subsequently released into the cytoplasm. As the RNP associated viral proteins contain nuclear localization signals (NLS), they are rapidly imported into the nucleus through the nuclear pore complexes [For review: (Palese and Shaw 2007; Wright et al. 2007)].

2.2.2 Genome Replication/Transcription

The IVs pursue a nuclear replication strategy. In order to accomplish nuclear import and export of its genome, the virus utilizes the cellular transport machinery. The transport of the RNPs into the nucleus is likely to be mediated by the viral PB2 and NP. They carry nuclear localization signals and interact with the α-class of karyopherin import receptor proteins (Neumann et al. 1997; Palese and Shaw 2007; Wang et al. 1997; Weber et al. 1998). Their nuclear import also depends on the presence of the import factors karyopherin α- and β, Ran and p10 (O'Neill et al. 1995). Furthermore, it was shown that the interaction of PB2 and NP with importin $\alpha 1$ is a host range determinant as adaptive mutations in both proteins enhance their binding to importin $\alpha 1$ and increase their transport into the nucleus of mammalian cells. In avian cells these effects were not observed (Gabriel et al. 2008).

The viral genomic segments are replicated and transcribed by the viral RdRp as part of the RNPs in the nucleus of the infected cell. The vRNA is directly transcribed to mRNA and, in addition, serves as a template for a complementary copy (cRNA), which itself is the template for new vRNA [For review: (Palese and Shaw 2007; Wright et al. 2007)]. PB1 is predicted to contain the central location of the polymerase domain. The PB2 subunit was identified as the site of cap binding. Various studies have shown that PB1-F2 has several effects: (i) it can induce apoptosis in a cell type-dependent manner; (ii) it is able to promote inflammation; and (iii) it can up-regulate viral polymerase activity by its interaction with the PB1 subunit. These properties could contribute to an enhanced pathogenicity. The underlying mechanisms are not fully understood and some effects of PB1-F2 might be strain- and host-specific (Mazur et al. 2008). The cap-snatching endonuclease of the IV RdRp resides in the PA subunit. By interaction of the C-termini of PA and PB1 with the *N*-termini of PB1 and PB2, respectively, the RdRp complex is formed. (Dias et al. 2009). The viral NP which associates with the genomic and anti-genomic viral RNAs is an essential cofactor of the viral replicative complex (Huang et al. 1990). Genomic IV RNAs carry at their 5′- and 3′-ends conserved nucleotide sequences of 13 and 12 bases, respectively. As these sequences are in part complementary, the ends of viral RNAs can engage in base-pairing interactions resulting in a partially double-stranded promoter structure (Fischer et al. 1998; Flick et al. 1996; Luo and Palese 1992). The viral RdRp binds to these promoter structures of the viral RNA segments and subsequently initiates RNA synthesis (Palese and Shaw 2007). In comparison to replication more is known about viral transcription. The RdRp does not synthesize the 5′-cap structures (m7GpppNm) that are needed for efficient transport and translation of the viral mRNAs. Instead, the 5′-cap structures from cellular polymerase II transcripts are transferred to viral mRNAs (Krug et al. 1979). It was realized quite early on that IV is inhibited under conditions where the cellular RNA polymerase II is blocked (Rott and Scholtissek 1970). The requirement for capped mRNA 5′-ends to function as primers in viral transcription most likely explains the dependency of viral replication on RNA polymerase II activity.

In the early phase of infection the viral genome is both transcribed and replicated at high rates. At later time points (>3.5 h p.i.) transcription decreases, whereas the replication of viral RNAs continues (Shapiro et al. 1987). It is not known in detail which regulatory factor(s) participate in this process. However, analysis of viral replication in vitro suggests that a fraction of the NP which is not associated with the viral RNPs may have a regulatory role in the switch from viral transcription to replication (Shapiro and Krug 1988). Furthermore, the viral M1 matrix protein that accumulates to high levels during the late phase of infection has been shown to inhibit viral transcription (Hankins et al. 1989; Perez and Donis 1998; Ye et al. 1987). M1 may therefore also be involved in the down regulation of viral mRNA synthesis late in infection.

2.2.3 Controlling the Cell

The NS segment encodes the NS1 protein, which is translated from unspliced mRNA, as well as the NEP protein, which is translated from spliced mRNA transcripts. The NS1 protein has been shown to be a major pathogenicity factor [For review: (Hale et al. 2008)]. As such, it can impair host innate and adaptive immunity in a number of ways. It can block the function of 2′-5′-oligoadenylate synthetase (OAS) (Bergmann et al. 2000) and bind to double-stranded RNA (dsRNA), thereby suppressing the activation of ds RNA-activated protein kinase (PKR), both important regulators of translation that can induce the host apoptotic response and type I interferon (IFN) production (Marjuki et al. 2007). Furthermore, it can inhibit retinoic acid-inducible gene I (RIG-I)-mediated induction of IFN by: (i) binding to RIG-1, and preventing it from binding to single-stranded RNA (ssRNA) bearing 5′-phosphates (Qian et al. 1994); or (ii) forming a NS1/RIG I-RNA complex (Falcon et al. 2004); or (iii) interacting with the ubiquitin ligase TRIM25 and inhibiting TRIM25-mediated RIG-I CARD ubiquitination (Gack et al. 2009). Another way by which NS1 impairs the production of IFN is to prevent the activation of transcription factors such as ATF-2/c-Jun, NF-κB, and IRF-3/5/7, all of which stimulate IFN production (Gambotto et al. 2008; Koopmans et al. 2004). By forming an inhibitory complex with NXF1/TAP, p15/NXT, Rae1/mrnp41, and E1B-AP5, which are important factors in the mRNA export machinery, NS1 decreases cellular mRNA transport in order to render cells highly permissive to IV replication (Robb et al. 2009). NS1 can also inhibit the 3′-end processing of cellular pre-mRNAs (including IFN pre-mRNA) through interaction with the cellular proteins CPSF30 (Treanor et al. 1989) and PABII (Bergmann et al. 2000).

Additionally, many studies have highlighted the importance of the interaction between the RNP complex and NS1 for viral replication (Donelan et al. 2003; Krug et al. 2003; Maines et al. 2005; Mibayashi et al. 2007). The NS1 protein was shown to interact with the RNP complex in vivo (Maines et al. 2005), and truncated NS1 affected production of vRNA, but not of cRNA and mRNA in infected MDCK cells, implicating NS1 in the regulation of replication (Donelan et al. 2003). The NS1 protein was also found to regulate the time course of viral RNA synthesis during infection as a mutant virus with two amino acid changes at positions 123 and 124 deregulated the normal time course of viral RNA synthesis (Mibayashi et al. 2007). Interestingly, characterization of H7-type HPAIV with reassorted NS segments from H5- and H7-type avian strains, generated by reverse genetics, demonstrated altered growth kinetics of the reassortant viruses that differed from the wild type. Surprisingly, the effects differed between cells of mammalian or avian origin; and molecular analysis revealed that the reassorted NS segments were not only responsible for alterations in the anti-viral host response, but furthermore affected viral genome replication and transcription as well as nuclear RNP export. IFN-beta expression and the induction of apoptosis were found to be inversely correlated with the magnitude of viral growth, while the NS allele, virus subtype and levels of NS1 protein expression showed no correlation. Thus, these results demonstrate that the origin of the NS segment can have a dramatic effect on the replication efficiency and host range of HPAIV. Overall, these data suggest that the propagation of NS reassortant IV is affected at multiple steps of the viral lifecycle as a result of the different activities of the NS1 protein on multiple viral and host functions (Wang et al. 2010). The NEP protein which interacts with the viral M1 protein and mediates the export of vRNPs from the nucleus to the cytoplasm (Nemeroff et al. 1998), has also been shown to play a role in the regulation of viral replication and transcription. However a direct interaction between NEP and the viral polymerase complex has not been demonstrated. Nevertheless, it is evident that both the NS1 and NEP proteins play important roles in viral pathogenicity and replication.

2.2.4 Assembly and Release

In the late phase of the replication cycle newly synthesized viral RNPs are exported to the cytoplasm, and two viral proteins have been suggested to play important roles in this transport event. It has been observed that the appearance of the viral M1 protein in the nucleus is required for the subsequent export of viral genomes, but the molecular basis of this dependency remains elusive (Subbarao et al. 1998). The M1 protein is known to associate with viral nucleocapsids and the NEP protein. It has been shown that the NEP protein contains a leucine-rich nuclear export signal by which it interacts with nucleoporins (Weber et al. 1998). In light of this finding, it was suggested that the M1-NEP complex mediates the export of associated viral nucleocapsids in the late phase of the replication. Previously, it was demonstrated that IV activates the Raf/MEK/ERK-signal transduction cascade, which is an essential pre-requisite for efficient nuclear RNP export (Ludwig et al. 2004; Pleschka et al. 2001). Surprisingly, other IV-induced events strongly contribute to the nuclear RNP export, namely the activation of NF-kB, as well as the apoptotic activity of caspase3 (Wurzer et al. 2003, 2004). As these cellular factors and mechanisms are essential for IV replication, their inhibition strongly affects IV replication. Because they are encoded in the host genome, the virus can hardly become resistant by mutation, making them ideal targets for therapeutic intervention (Ludwig et al. 2003). In the cytoplasm, the M1 protein inhibits re-import of the nucleocapsids into the nucleus, possibly by masking the karyophilic signals of the NP (Bui et al. 1996). Another mechanism of cytoplasmic retention of the RNPs was proposed to be due to the ability of NP to associate with the actin cytoskeleton (Avalos et al. 1997; Digard et al. 1999).

The mature HA and NA glycoproteins and the nonglycosylated M2 are finally integrated into the plasma membrane as trimers (HA) or tetramers (NA, M2), respectively. M1 assembles in patches at the cell membrane. It is thought to associate with the glycoproteins (HA and NA) and to recruit the RNPs to the plasma membrane in the late phase of the replication cycle. Finally, the viral RNPs become enveloped by a cellular bi-lipid layer carrying the HA, NA, and M2 proteins resulting in budding of new virus particles from the apical cell surface. Interestingly, HA membrane accumulation actually triggers the essential ERK signaling. This represents an auto-regulative feedback loop that assures nuclear RNP export at a time point when all other viral components are ready for budding (Marjuki 2006). Furthermore, higher polymerase activity of a human IV enhances activation of the HA-induced Raf/MEK/ERK-signal cascade resulting in more efficient nuclear RNP-export as well as virus production (Marjuki et al. 2007).

By the receptor destroying neuraminidase activity of the NA, the progeny virions are able to detach from the cell surface, to which they would otherwise reattach by the HA activity. As such the NA, which is a type II glycoprotein that follows the exocytotic transport pathway and presumably encounters many of the same enzymes that HA does, plays an essential role in release and spread of progeny virions. Moreover, NA was also shown to be important for the initiation of IV infection in human airway epithelium (Matrosovich et al. 2004). A functional balance between HA and NA in IV infections seems to be highly important. Both proteins recognize sialic acid. HA binds to sialic acid-containing receptors on target cells to initiate virus infection, whereas NA cleaves sialic acids from cellular receptors to facilitate progeny virus release. Studies have revealed that an optimal interplay between these receptor-binding and receptor-destroying activities of HA and NA is required for efficient virus replication. An existing balance between the antagonistic HA and NA functions can be disturbed by reassortment, virus transmission to a new host, or therapeutic inhibition of NA. The resulting decrease in the viral replicative fitness can be overcome by restoration of the functional balance via compensatory mutations in HA, NA or both proteins (Wagner et al. 2002).

2.3 Antigenic Drift and Antigenic Shift

The polymerase complex of IV does not possess a proof reading activity, thus numerous mutations accumulate in the viral genome during ongoing replication (Palese and Shaw 2007) leading to changes in all proteins. This includes conformational alteration of HA- and NA-epitopes against which neutralizing antibodies are generated. Influenza A viruses are categorized by antigenic differences of the HA and NA-proteins. The high mutation rate combined with the high replication rate results in a multitude of new variants produced in each replication cycle, thus allowing the virus to rapidly adapt to changes in the environment. This results in an escape of the existing immunity and in resistance to drugs acting directly against viral functions. Gradual changes of the antigenic properties that make existing vaccines less or noneffective are described as *antigenic drift* and demand for new compositions of the yearly vaccines.

Due to the nature of their segmented genome, IV can independently recombine segments upon the infection of a cell with two different viruses. This is described as *genetic reassortment*. Through its receptor binding and fusion activity as well as its antigenic properties, the HA is a major determinant of tissue tropism, viral spread, and pathogenicity in IV-infected organisms. Today 17 HA-subtypes (H1–H17) and 9 NA-subtypes (N1–N9) are known, which can mix and lead to new antigenic properties (Palese and Shaw 2007; Webster et al. 1992; Wright et al. 2007). Not all combination will ultimately be advantageous, but can lead to the generation of a virus that combines the ability to replicate in humans with novel antigenic properties (*antigenic shift*). This has happened at least three times in the last century, resulting in the pandemics of 1918 ("Spanish Flu"), 1957 ("Asian Flu") and 1968 ("Hong Kong Flu") that together caused up to 40 million deaths. The 2009 introduction of the new pandemic H1N1-type swine origin IV (S-OIV) into the human population, which comprises a reassortant IV harboring segments of human and avian IV, as well as of swine IV belonging to the North American and Eurasian lineage demonstrates that the question is not "if" but "when" will such new pandemics occur (Horimoto and Kawaoka 2001; Webby and Webster 2003; Webster 1997).

Besides pandemic variants that can occur when human and avian IV reassort in porcine hosts (regarded as "mixing vessels") (Webster et al. 1995; Webster 1997a, b), HPAIV strains have directly infected humans, as happened in Hong Kong in 1997 (Claas et al. 1998; de Jong et al. 1997; Subbarao et al. 1998), and thereafter during vast outbreaks of avian influenza (Fouchier et al. 2004; Koopmans et al. 2004). These viruses show an extremely high virulence in humans with case fatality rates up to 60 % (World Health Organization 2005).

2.4 "Flu"- the Disease

The virus that usually causes a respiratory disease in humans [For references: (Wilschut 2005)] is transmitted by aerosol droplets and contaminated hands and can already be shed before the onset of symptoms (Cox et al. 2004). Therefore, high population density and dry air leading to reduced protection of respiratory epithelium by the mucus are conditions that promote transmission of the virus.

The infection with IV in humans is normally limited to the respiratory tract. Here, proteases released in the epithelium are present that activate the HA to allow further infections (Sect. 2.2.1) [For review (Ludwig et al. 1999)]. Innate immunity as well as the adaptive immune system will normally restrict virus propagation. Therefore, population groups which have a less protective immune system, such as young children up to 2 years and older persons over 65 as well as immunocompromised or chronically diseased persons, are especially at risk. The replication of the virus leads to the lysis of the epithelial cells and enhanced mucus production, causing runny nose and cough. Also, inflammation and edema at the replication site due to cytokines released contribute to the disease. This can lead to fever and related symptoms. Bacterial superinfections of the harmed tissue can further complicate the situation. Normally, onset of systemic (fever, myaglia, headaches, and severe malaise) and respiratory (coughing, sore throat, and rhinitis) symptoms occurs after about 2 days incubation period and can last for about 7 to 10 days. Coughing and overall weakness can persist for up to 2 weeks. If the virus spreads from the bronchiolar tract to the alveoli, viral pneumonia and interstitial pneumonitis with mononuclear and hemorrhage infiltration and finally lysis of the inte-alveolar space are all possible (Wilschut 2005).

This scenario is a likely picture in case of infection with a pandemic IV strain, where the individual has not had a prior exposure to the virus and the innate immunity reaction can lead to a strong immune reaction. High virus replication will induce secretion of large quantities of cytokines by the infected epithelia and will stimulate inflammatory processes. Together with the destruction of the epithelia, this results in an influx of fluids into the alveoli leading to hypoxia and acute respiratory distress syndrome that may cause the death within a short period of time (1–2 days after onset). This scenario might also be caused by additional viral factors enhancing pathogenicity. Such factors that are not yet well defined but probably contributed to the devastating outcome of the "Spanish Flu" (Wilschut 2005).

Accurate and rapid diagnosis of the disease is essential for effective treatment, especially with anti-viral substances, as virus replication and therefore illness progresses rapidly. Samples can be tested serologically, by cell culture or RT-PCR for strain typing and should be done within days after onset of symptoms (Wilschut 2005).

Since pandemic virus strains usually possess unique antigenic characteristics, current vaccines will be ineffective once such a virus emerges. Regarding the vast possibilities for such a novel strain to "travel" around the world (Hufnagel et al. 2004), it becomes evident that effective countermeasures are required for the fight against these foes. In recent outbreaks of avian viruses that infected humans (Chen et al. 2004; Hatta and Kawaoka 2002; Li et al. 2004) a mortality rate of about 60 % was observed (World Health Organization 2005). Fortunately, until now these particular viruses have not acquired the ability to spread in the human population. However, any novel virus strain emerging in the future may have such a capability (Webby and Webster 2003).

As every virus depends on its host cell, cellular functions essential for viral replication may also be suitable targets for anti-viral therapy. In this respect, intracellular signaling cascades activated by the virus, in particular MAPK pathways, have recently come into focus (see Sect. 2.2.4) (Ludwig et al. 1999, 2003).

References

Avalos RT, Yu Z, Nayak DP (1997) Association of influenza virus NP and M1 proteins with cellular cytoskeletal elements in influenza virus-infected cells. J Virol 71:2947–2958

Bergmann M, Garcia-Sastre A, Carnero E, Pehamberger H, Wolff K, Palese P, Muster T (2000) Influenza virus NS1 protein counteracts PKR-mediated inhibition of replication. J Virol 74:6203–6206

Bottcher E, Matrosovich T, Beyerle M, Klenk HD, Garten W, Matrosovich M (2006) Proteolytic activation of influenza viruses by serine proteases TMPRSS2 and HAT from human airway epithelium. J Virol 80:9896–9898

Braakman I, Hoover-Litty H, Wagner KR, Helenius A (1991) Folding of influenza hemagglutinin in the endoplasmic reticulum. J Cell Biol 114:401–411

Bui M, Whittaker G, Helenius A (1996) Effect of M1 protein and low pH on nuclear transport of influenza virus ribonucleoproteins. J Virol 70:8391–8401

Ceriotti A, Colman A (1990) Trimer formation determines the rate of influenza virus haemagglutinin transport in the early stages of secretion in Xenopus oocytes. J Cell Biol 111:409–420

Chen W, Helenius J, Braakman I, Helenius A (1995) Cotranslational folding and calnexin binding during glycoprotein synthesis. Proc Natl Acad Sci U S A 92:6229–6233

Chen W, Calvo PA, Malide D, Gibbs J, Schubert U, Bacik I, Basta S, O'Neill R, Schickli J, Palese P, Henklein P, Bennink JR, Yewdell JW (2001) A novel influenza A virus mitochondrial protein that induces cell death. Nat Med 7:1306–1312

Chen H, Deng G, Li Z, Tian G, Li Y, Jiao P, Zhang L, Liu Z, Webster RG, Yu K (2004) The evolution of H5N1 influenza viruses in ducks in southern China. Proc Natl Acad Sci U S A 101:10452–10457

Claas EC, de Jong JC, van Beek R, Rimmelzwaan GF, Osterhaus AD (1998) Human influenza virus A/HongKong/156/97 (H5N1) infection. Vaccine 16:977–978

Connor RJ, Kawaoka Y, Webster RG, Paulson JC (1994) Receptor specificity in human, avian, and equine H2 and H3 influenza virus isolates. Virology 205:17–23

Copeland CS, Doms RW, Bolzau EM, Webster RG, Helenius A (1986) Assembly of influenza hemagglutinin trimers and its role in intracellular transport. J Cell Biol 103:1179–1191

Cox RJ, Brokstad KA, Ogra P (2004) Influenza virus: immunity and vaccination strategies. Comparison of the immune response to inactivated and live, attenuated influenza vaccines. Scand J Immunol 59:1–15

de Jong JC, Claas EC, Osterhaus AD, Webster RG, Lim WL (1997) A pandemic warning? Nature 389:554

Dias A, Bouvier D, Crepin T, McCarthy AA, Hart DJ, Baudin F, Cusack S, Ruigrok RW (2009) The cap-snatching endonuclease of influenza virus polymerase resides in the PA subunit. Nature 458:914–918

Digard P, Elton D, Bishop K, Medcalf E, Weeds A, Pope B (1999) Modulation of nuclear localization of the influenza virus nucleoprotein through interaction with actin filaments. J Virol 73:2222–2231

Donelan NR, Basler CF, Garcia-Sastre A (2003) A recombinant influenza A virus expressing an RNA-binding-defective NS1 protein induces high levels of beta interferon and is attenuated in mice. J Virol 77:13257–13266

Drummer HE, Jackson DC, Brown LE (1993) Modulation of CD4+ T-cell recognition of influenza hemagglutinin by carbohydrate side chains located outside a T-cell determinant. Virology 192:282–289

Eierhoff T, Hrincius ER, Rescher U, Ludwig S, Ehrhardt C (2010) The epidermal growth factor receptor (EGFR) promotes uptake of influenza A viruses (IAV) into host cells. PLoS Pathog 6:1–16

Falcon AM, Marion RM, Zurcher T, Gomez P, Portela A, Nieto A, Ortin J (2004) Defective RNA replication and late gene expression in temperature-sensitive influenza viruses expressing deleted forms of the NS1 protein. J Virol 78:3880–3888

Fischer C, Schroth-Diez B, Herrmann A, Garten W, Klenk HD (1998) Acylation of the influenza hemagglutinin modulates fusion activity. Virology 248:284–294

Flick R, Neumann G, Hoffmann E, Neumeier E, Hobom G (1996) Promoter elements in the influenza vRNA terminal structure. RNA 2:1046–1057

Fouchier RA, Schneeberger PM, Rozendaal FW, Broekman JM, Kemink SA, Munster V, Kuiken T, Rimmelzwaan GF, Schutten M, Van Doornum GJ, Koch G, Bosman A, Koopmans M, Osterhaus AD (2004) Avian influenza A virus (H7N7) associated with human conjunctivitis and a fatal case of acute respiratory distress syndrome. Proc Natl Acad Sci U S A 101:1356–1361

Gabriel G, Herwig A, Klenk HD (2008) Interaction of polymerase subunit PB2 and NP with importin alpha1 is a determinant of host range of influenza A virus. PLoS Pathog 4:e11

Gack MU, Albrecht RA, Urano T, Inn KS, Huang IC, Carnero E, Farzan M, Inoue S, Jung JU, Garcia-Sastre A (2009) Influenza A virus NS1 targets the ubiquitin ligase TRIM25 to evade recognition by the host viral RNA sensor RIG-I. Cell Host Microbe 5:439–449

Gallagher P, Henneberry J, Wilson I, Sambrook J, Gething MJ (1988) Addition of carbohydrate side chains at novel sites on influenza virus hemagglutinin can modulate the folding, transport, and activity of the molecule. J Cell Biol 107:2059–2073

Gambaryan AS, Marinina VP, Tuzikov AB, Bovin NV, Rudneva IA, Sinitsyn BV, Shilov AA, Matrosovich MN (1998) Effects of host-dependent glycosylation of hemagglutinin on receptor- binding properties on H1N1 human influenza A virus grown in MDCK cells and in embryonated eggs. Virology 247:170–177

Gambotto A, Barratt-Boyes SM, de Jong MD, Neumann G, Kawaoka Y (2008) Human infection with highly pathogenic H5N1 influenza virus. Lancet 371:1464–1475

Gething MJ, McCammon K, Sambrook J (1986) Expression of wild-type and mutant forms of influenza hemagglutinin: the role of folding in intracellular transport. Cell 46:939–950

Gotoh B, Ogasawara T, Toyoda T, Inocencio NM, Hamaguchi M, Nagai Y (1990) An endoprotease homologous to the blood clotting factor X as a determinant of viral tropism in chick embryo. EMBO J 9:4189–4195

Gottlieb TA, Gonzalez A, Rizzolo L, Rindler MJ, Adesnik M, Sabatini DD (1986) Sorting and endocytosis of viral glycoproteins in transfected polarized epithelial cells. J Cell Biol 102:1242–1255

Hale BG, Randall RE, Ortin J, Jackson D (2008) The multifunctional NS1 protein of influenza A viruses. J Gen Virol 89:2359–2376

Hankins RW, Nagata K, Bucher DJ, Popple S, Ishihama A (1989) Monoclonal antibody analysis of influenza virus matrix protein epitopes involved in transcription inhibition. Virus Genes 3:111–126

Hatta M, Kawaoka Y (2002) The continued pandemic threat posed by avian influenza viruses in Hong Kong. Trends Microbiol 10:340–344

Hebert DN, Foellmer B, Helenius A (1995) Glucose trimming and reglucosylation determine glycoprotein association with calnexin in the endoplasmic reticulum. Cell 81:425–433

Hebert DN, Zhang JX, Chen W, Foellmer B, Helenius A (1997) The number and location of glycans on influenza hemagglutinin determine folding and association with calnexin and calreticulin. J Cell Biol 139:613–623

Horimoto T, Kawaoka Y (2001) Pandemic threat posed by avian influenza A viruses. Clin Microbiol Rev 14:129–149

Horimoto T, Nakayama K, Smeekens SP, Kawaoka Y (1994) Proprotein-processing endoproteases PC6 and furin both activate hemagglutinin of virulent avian influenza viruses. J Virol 68:6074–6078

Huang TS, Palese P, Krystal M (1990) Determination of influenza virus proteins required for genome replication. J Virol 64:5669–5673

Hufnagel L, Brockmann D, Geisel T (2004) Forecast and control of epidemics in a globalized world. Proc Natl Acad Sci U S A 101:15124–15129

Hurtley SM, Bole DG, Hoover-Litty H, Helenius A, Copeland CS (1989) Interactions of misfolded influenza virus hemagglutinin with binding protein (BiP). J Cell Biol 108:2117–2126

Kawaoka Y, Naeve CW, Webster RG (1984) Is virulence of H5N2 influenza viruses in chickens associated with loss of carbohydrate from the hemagglutinin? Virology 139:303–316

Kido H, Yokogoshi Y, Sakai K, Tashiro M, Kishino Y, Fukutomi A, Katunuma N (1992) Isolation and characterization of a novel trypsin-like protease found in rat bronchiolar epithelial Clara cells. A possible activator of the viral fusion glycoprotein. J Biol Chem 267:13573–13579

Kido H, Sakai K, Kishino Y, Tashiro M (1993) Pulmonary surfactant is a potential endogenous inhibitor of proteolytic activation of Sendai virus and influenza A virus. FEBS Lett 322:115–119

Klenk HD, Garten W (1994) Host cell proteases controlling virus pathogenicity. Trends Microbiol 2:39–43

Klenk HD, Rott R, Orlich M, Blodorn J (1975) Activation of influenza A viruses by trypsin treatment. Virology 68:426–439

Koopmans M, Wilbrink B, Conyn M, Natrop G, van der Nat H, Vennema H, Meijer A, van Steenbergen J, Fouchier R, Osterhaus A, Bosman A (2004) Transmission of H7N7 avian influenza A virus to human beings during a large outbreak in commercial poultry farms in the Netherlands. Lancet 363:587–593

Krug RM, Broni BA, Bouloy M (1979) Are the 5′ ends of influenza viral mRNAs synthesized in vivo donated by host mRNAs? Cell 18:329–334

Krug RM, Yuan W, Noah DL, Latham AG (2003) Intracellular warfare between human influenza viruses and human cells: the roles of the viral NS1 protein. Virology 309:181–189

Lambrecht B, Schmidt MF (1986) Membrane fusion induced by influenza virus hemagglutinin requires protein bound fatty acids. FEBS Lett 202:127–132

Lazarowitz SG, Choppin PW (1975) Enhancement of the infectivity of influenza A and B viruses by proteolytic cleavage of the hemagglutinin polypeptide. Virology 68:440–454

Li KS, Guan Y, Wang J, Smith GJ, Xu KM, Duan L, Rahardjo AP, Puthavathana P, Buranathai C, Nguyen TD, Estoepangestie AT, Chaisingh A, Auewarakul P, Long HT, Hanh NT, Webby RJ, Poon LL, Chen H, Shortridge KF, Yuen KY, Webster RG, Peiris JS (2004) Genesis of a highly pathogenic and potentially pandemic H5N1 influenza virus in eastern Asia. Nature 430: 209–213

Ludwig S, Pleschka S, Wolff T (1999) A fatal relationship–influenza virus interactions with the host cell. Viral Immunol 12:175–196

Ludwig S, Planz O, Pleschka S, Wolff T (2003) Influenza-virus-induced signaling cascades: targets for antiviral therapy? Trends Mol Med 9:46–52

Ludwig S, Wolff T, Ehrhardt C, Wurzer WJ, Reinhardt J, Planz O, Pleschka S (2004) MEK inhibition impairs influenza B virus propagation without emergence of resistant variants. FEBS Lett 561:37–43

Luo G, Palese P (1992) Genetic analysis of influenza virus. Curr Opin Genet Dev 2:77–81
Ma W, Brenner D, Wang Z, Dauber B, Ehrhardt C, Hogner K, Herold S, Ludwig S, Wolff T, Yu K, Richt JA, Planz O, Pleschka S (2010) The NS segment of an H5N1 highly pathogenic avian influenza virus (HPAIV) is sufficient to alter replication efficiency, cell tropism, and host range of an H7N1 HPAIV. J Virol 84:2122–2133
Maines TR, Lu XH, Erb SM, Edwards L, Guarner J, Greer PW, Nguyen DC, Szretter KJ, Chen LM, Thawatsupha P, Chittaganpitch M, Waicharoen S, Nguyen DT, Nguyen T, Nguyen HH, Kim JH, Hoang LT, Kang C, Phuong LS, Lim W, Zaki S, Donis RO, Cox NJ, Katz JM, Tumpey TM (2005) Avian influenza (H5N1) viruses isolated from humans in Asia in 2004 exhibit increased virulence in mammals. J Virol 79:11788–11800
Marjuki H, Alam MI, Ehrhardt C, Wagner R, Planz O, Klenk HD, Ludwig S, Pleschka S (2006) Membrane accumulation of influenza a virus hemagglutinin triggers nuclear export of the viral genome via PKCalpha mediated activation of ERK signaling. J Biol Chem 16707–16715
Marjuki H, Yen HL, Franks J, Webster RG, Pleschka S, Hoffmann E (2007) Higher polymerase activity of a human influenza virus enhances activation of the hemagglutinin-induced Raf/MEK/ERK signal cascade. Virol J 4:134
Marjuki H, Gornitzky A, Marathe BM, Ilyushina NA, Aldridge JR, Desai G, Webby RJ, Webster RG (2010) Influenza A virus-induced early activation of ERK and PI3 K mediates V-ATPase-dependent intracellular pH change required for fusion. Cell Microbiol 587–601
Matrosovich MN, Matrosovich TY, Gray T, Roberts NA, Klenk HD (2004) Neuraminidase is important for the initiation of influenza virus infection in human airway epithelium. J Virol 78:12665–12667
Mazur I, Anhlan D, Mitzner D, Wixler L, Schubert U, Ludwig S (2008) The proapoptotic influenza A virus protein PB1-F2 regulates viral polymerase activity by interaction with the PB1 protein. Cell Microbiol 10:1140–1152
Melikyan GB, Jin H, Lamb RA, Cohen FS (1997) The role of the cytoplasmic tail region of influenza virus hemagglutinin in formation and growth of fusion pores. Virology 235:118–128
Mibayashi M, Martinez-Sobrido L, Loo YM, Cardenas WB, Gale M Jr, Garcia-Sastre A (2007) Inhibition of retinoic acid-inducible gene I-mediated induction of beta interferon by the NS1 protein of influenza A virus. J Virol 81:514–524
Mir-Shekari SY, Ashford DA, Harvey DJ, Dwek RA, Schulze IT (1997) The glycosylation of the influenza A virus hemagglutinin by mammalian cells. A site-specific study. J Biol Chem 272:4027–4036
Munk K, Pritzer E, Kretzschmar E, Gutte B, Garten W, Klenk HD (1992) Carbohydrate masking of an antigenic epitope of influenza virus haemagglutinin independent of oligosaccharide size. Glycobiology 2:233–240
Naeve CW, Williams D (1990) Fatty acids on the A/Japan/305/57 influenza virus hemagglutinin have a role in membrane fusion. EMBO J 9:3857–3866
Nemeroff ME, Barabino SM, Li Y, Keller W, Krug RM (1998) Influenza virus NS1 protein interacts with the cellular 30 kDa subunit of CPSF and inhibits 3′end formation of cellular pre-mRNAs. Mol Cell 1:991–1000
Neumann G, Castrucci MR, Kawaoka Y (1997) Nuclear import and export of influenza virus nucleoprotein. J Virol 71:9690–9700
Ohuchi M, Ohuchi R, Feldmann A, Klenk HD (1997a) Regulation of receptor binding affinity of influenza virus hemagglutinin by its carbohydrate moiety. J Virol 71:8377–8384
Ohuchi R, Ohuchi M, Garten W, Klenk HD (1997b) Oligosaccharides in the stem region maintain the influenza virus hemagglutinin in the metastable form required for fusion activity. J Virol 71:3719–3725
Okumura Y, Takahashi E, Yano M, Ohuchi M, Daidoji T, Nakaya T, Bottcher E, Garten W, Klenk HD, Kido H (2010) Novel type II transmembrane serine proteases, MSPL and TMPRSS13, Proteolytically activate membrane fusion activity of the hemagglutinin of highly pathogenic avian influenza viruses and induce their multicycle replication. J Virol 84: 5089–5096

O'Neill RE, Jaskunas R, Blobel G, Palese P, Moroianu J (1995) Nuclear import of influenza virus RNA can be mediated by viral nucleoprotein and transport factors required for protein import. J Biol Chem 270:22701–22704

Palese P, Shaw ML (2007) Orthomyxoviridae: the virus and their replication. In: Knipe DM, Howley PM (eds) Fields—Virology, 5th edn. Lippincott Williams & Wilkins, Philadelphia, pp 1647–1689

Paulson J (ed) (1985) Interactions of animal viruses with cell surface receptors, vol 2. Academic Press, Orlando

Perez DR, Donis RO (1998) The matrix 1 protein of influenza A virus inhibits the transcriptase activity of a model influenza reporter genome in vivo. Virology 249:52–61

Peterson JR, Ora A, Van PN, Helenius A (1995) Transient, lectin-like association of calreticulin with folding intermediates of cellular and viral glycoproteins. Mol Biol Cell 6:1173–1184

Philipp HC, Schroth B, Veit M, Krumbiegel M, Herrmann A, Schmidt MF (1995) Assessment of fusogenic properties of influenza virus hemagglutinin deacylated by site-directed mutagenesis and hydroxylamine treatment. Virology 210:20–28

Pleschka S, Wolff T, Ehrhardt C, Hobom G, Planz O, Rapp UR, Ludwig S (2001) Influenza virus propagation is impaired by inhibition of the Raf/MEK/ERK signalling cascade. Nat Cell Biol 3:301–305

Qian XY, Alonso-Caplen F, Krug RM (1994) Two functional domains of the influenza virus NS1 protein are required for regulation of nuclear export of mRNA. J Virol 68:2433–2441

Rindler MJ, Ivanov IE, Plesken H, Rodriguez-Boulan E, Sabatini DD (1984) Viral glycoproteins destined for apical or basolateral plasma membrane domains traverse the same Golgi apparatus during their intracellular transport in doubly infected Madin-Darby canine kidney cells. J Cell Biol 98:1304–1319

Robb NC, Smith M, Vreede FT, Fodor E (2009) NS2/NEP protein regulates transcription and replication of the influenza virus RNA genome. J Gen Virol 90:1398–1407

Rodriguez-Boulan E, Paskiet KT, Sabatini DD (1983) Assembly of enveloped viruses in Madin-Darby canine kidney cells: polarized budding from single attached cells and from clusters of cells in suspension. J Cell Biol 96:866–874

Rodriguez-Boulan E, Paskiet KT, Salas PJ, Bard E (1984) Intracellular transport of influenza virus hemagglutinin to the apical surface of Madin-Darby canine kidney cells. J Cell Biol 98:308–319

Rott R, Scholtissek C (1970) Specific inhibition of influenza replication by alpha-amanitin. Nature 228:56

Rott R, Klenk HD, Nagai Y, Tashiro M (1995) Influenza viruses, cell enzymes, and pathogenicity. Am J Respir Crit Care Med 152:S16–S19

Sakai K, Kohri T, Tashiro M, Kishino Y, Kido H (1994) Sendai virus infection changes the subcellular localization of tryptase Clara in rat bronchiolar epithelial cells. Eur Respir J 7:686–692

Scheiblauer H, Reinacher M, Tashiro M, Rott R (1992) Interactions between bacteria and influenza A virus in the development of influenza pneumonia. J Infect Dis 166:783–791

Schmidt MF (1982) Acylation of viral spike glycoproteins: a feature of enveloped RNA viruses. Virology 116:327–338

Schulze IT (1997) Effects of glycosylation on the properties and functions of influenza virus hemagglutinin. J Infect Dis 176(Suppl 1):S24–S28

Shapiro GI, Krug RM (1988) Influenza virus RNA replication in vitro: synthesis of viral template RNAs and virion RNAs in the absence of an added primer. J Virol 62:2285–2290

Shapiro GI, Gurney T Jr, Krug RM (1987) Influenza virus gene expression: control mechanisms at early and late times of infection and nuclear-cytoplasmic transport of virus-specific RNAs. J Virol 61:764–773

Simpson DA, Lamb RA (1992) Alterations to influenza virus hemagglutinin cytoplasmic tail modulate virus infectivity. J Virol 66:790–803

Steinhauer DA, Wharton SA, Wiley DC, Skehel JJ (1991) Deacylation of the hemagglutinin of influenza A/Aichi/2/68 has no effect on membrane fusion properties. Virology 184:445–448

Stieneke-Grober A, Vey M, Angliker H, Shaw E, Thomas G, Roberts C, Klenk HD, Garten W (1992) Influenza virus hemagglutinin with multibasic cleavage site is activated by furin, a subtilisin-like endoprotease. EMBO J 11:2407–2414

Subbarao K, Klimov A, Katz J, Regnery H, Lim W, Hall H, Perdue M, Swayne D, Bender C, Huang J, Hemphill M, Rowe T, Shaw M, Xu X, Fukuda K, Cox N (1998) Characterization of an avian influenza A (H5N1) virus isolated from a child with a fatal respiratory illness. Science 279:393–396

Tashiro M, Ciborowski P, Reinacher M, Pulverer G, Klenk HD, Rott R (1987) Synergistic role of staphylococcal proteases in the induction of influenza virus pathogenicity. Virology 157: 421–430

Tashiro M, Yokogoshi Y, Tobita K, Seto JT, Rott R, Kido H (1992) Tryptase Clara, an activating protease for Sendai virus in rat lungs, is involved in pneumopathogenicity. J Virol 66: 7211–7216

Tatu U, Helenius A (1997) Interactions between newly synthesized glycoproteins, calnexin and a network of resident chaperones in the endoplasmic reticulum. J Cell Biol 136:555–565

Taubenberger JK, Reid AH, Fanning TG (2000) The 1918 influenza virus: a killer comes into view. Virology 274:241–245

Treanor JJ, Snyder MH, London WT, Murphy BR (1989) The B allele of the NS gene of avian influenza viruses, but not the A allele, attenuates a human influenza A virus for squirrel monkeys. Virology 171:1–9

Veit M, Kretzschmar E, Kuroda K, Garten W, Schmidt MF, Klenk HD, Rott R (1991) Site-specific mutagenesis identifies three cysteine residues in the cytoplasmic tail as acylation sites of influenza virus hemagglutinin. J Virol 65:2491–2500

Wagner R, Matrosovich M, Klenk HD (2002) Functional balance between haemagglutinin and neuraminidase in influenza virus infections. Rev Med Virol 12:159–166

Wagner R, Herwig A, Azzouz N, Klenk HD (2005) Acylation-mediated membrane anchoring of avian influenza virus hemagglutinin is essential for fusion pore formation and virus infectivity. J Virol 79:6449–6458

Walker JA, Molloy SS, Thomas G, Sakaguchi T, Yoshida T, Chambers TM, Kawaoka Y (1994) Sequence specificity of furin, a proprotein-processing endoprotease, for the hemagglutinin of a virulent avian influenza virus. J Virol 68:1213–1218

Wang P, Palese P, O'Neill RE (1997) The NPI-1/NPI-3 (karyopherin alpha) binding site on the influenza a virus nucleoprotein NP is a nonconventional nuclear localization signal. J Virol 71:1850–1856

Wang Z, Robb NC, Lenz E, Wolff T, Fodor E, Pleschka S (2010) NS reassortment of an H7-type highly pathogenic avian influenza virus affects its propagation by altering the regulation of viral RNA production and antiviral host response. J Virol 84:11323–11335

Webby RJ, Webster RG (2003) Are we ready for pandemic influenza? Science 302:1519–1522

Weber F, Kochs G, Gruber S, Haller O (1998) A classical bipartite nuclear localization signal on Thogoto and influenza A virus nucleoproteins. Virology 250:9–18

Webster RG (1997a) Influenza virus: transmission between species and relevance to emergence of the next human pandemic. Arch Virol Suppl 13:105–113

Webster RG (1997b) Predictions for future human influenza pandemics. J Infect Dis 176 (Suppl 1):S14–S19

Webster RG (1999) 1918 Spanish influenza: the secrets remain elusive. Proc Natl Acad Sci U S A 96:1164–1166

Webster RG, Bean WJ, Gorman OT, Chambers TM, Kawaoka Y (1992) Evolution and ecology of influenza A viruses. Microbiol Rev 56:152–179

Webster RG, Sharp GB, Claas EC (1995) Interspecies transmission of influenza viruses. Am J Respir Crit Care Med 152:S25–S30

Wilschut J, McElhaney JE (2005) Influenza. Mosby Elsvier Limited, Spain

World Health Organization (2005) Communicable disease surveillance & response (CSR). Avian Influenza. [Online.]

Wright PF, Naumann G, Kawaoka Y (2007) Orthomyxoviruses. In: Knipe DM, Howley PM (eds) Fields—virology, 5th edn. Lippincott Williams & Wilkins, Philadelphia, pp 1691–1740

Wurzer WJ, Planz O, Ehrhardt C, Giner M, Silberzahn T, Pleschka S, Ludwig S (2003) Caspase 3 activation is essential for efficient influenza virus propagation. EMBO J 22:2717–2728

Wurzer WJ, Ehrhardt C, Pleschka S, Berberich-Siebelt F, Wolff T, Walczak H, Planz O, Ludwig S (2004) NF-kappaB-dependent induction of tumor necrosis factor-related apoptosis-inducing ligand (TRAIL) and Fas/FasL is crucial for efficient influenza virus propagation. J Biol Chem 279:30931–30937

Ye ZP, Pal R, Fox JW, Wagner RR (1987) Functional and antigenic domains of the matrix (M1) protein of influenza A virus. J Virol 61:239–246

History of Swine Influenza

Stacey Schultz-Cherry, Christopher W. Olsen and Bernard C. Easterday

Abstract Swine influenza is a continual problem for the swine industry and can pose a public health threat as evidenced by the 2009 H1N1 influenza virus pandemic. Given its importance, it is not surprising to find papers describing the disease from the early 20th century. In this chapter, we discuss the history of swine influenza, the important role swine influenza virus has played in our understanding of influenza virus pathogenesis and virology, and its impact on public health worldwide.

Contents

Swine influenza (SI) has been a commonly recognized disease of swine for more than 90 years. Despite considerable interest and research efforts over the past 50 years, SI continues to be an important economic issue in swine production in diverse parts of the world today. Given the 2009 H1N1 pandemic, which was unfortunately known as "swine influenza" or "swine flu", it seems appropriate to review the origin of this important disease of swine, its importance in the swine production industry, and its relationship to influenzas of humans and other animals.

S. Schultz-Cherry (✉)
Department of Infectious Diseases, St. Jude Children's Research Hospital, Memphis, TN 38105, USA
e-mail: stacey.schultz-cherry@stjude.org

C. W. Olsen · B. C. Easterday
School of Veterinary of Veterinary Medicine,
University of Wisconsin-Madison, Madison, WI 53706, USA

Current Topics in Microbiology and Immunology (2011) 370: 21–27
DOI: 10.1007/82_2011_197

Published Online: 21 January 2012

Most reviews on SI will begin with some historical descriptions of when and where the disease was first observed. The earliest papers are from the early twentieth century. Koen (1919) stated that "Than the differential diagnosis of swine diseases there is no more important subject confronting the veterinary profession today" in "A Practical Method for the Field Diagnosis of Swine Diseases". Koen was a Bureau of Animal Industry inspector (USDA), in charge of hog cholera control work in Iowa. In that capacity he had abundant opportunity to investigate a variety of diseases of swine. He emphasized that his approach was "practical" diagnosis that was based on three things: history, clinical signs, and post-mortem lesions. He described the differential diagnosis of five conditions including "flu". Remarkably, his descriptions of "flu" in swine would match the appearance of any current SI natural outbreak. His comments in his closing arguments bring an intriguing perspective on the 1918 pandemic. He wrote that "Last fall and winter [1918] we were confronted with a new condition, if not a new disease. I believe I have as much to support this diagnosis in pigs as the physicians have to support a similar diagnosis in man. The similarity of the epidemic among people and the epizootic among pigs was so close, the reports so frequent, that an outbreak in the family would be followed immediately by an outbreak among the hogs, and vice versa, as to present a most striking coincidence if not suggesting a close relation between the two conditions. It looked like 'flu', it presented the identical symptoms of 'flu', it terminated like 'flu' and until proved it was not 'flu', I shall stand by that diagnosis". However, a 1921 report by Dr. Charles Murray, Professor of Veterinary Medicine, Iowa State College, presented an opposing view in an article entitled "WHAT IS HOG 'FLU'" (Murray 1921). He wrote that the disease in swine "...was unfortunately given the name of 'hog flu', a name which caused much apprehension among the agricultural people who were led to believe through the similarity of names that the diseases were the same cause and that the one affecting swine was transmissible to man. Such was not the case, and no such transmission ever occurred."He also stated that "So called 'swine flu', a name which, while it became quite popular through its association with the human disease, is nevertheless a misnomer...".

The issue of calling the disease "swine flu" led to continued debate at the annual meeting of the American Veterinary Medical Association in 1922 where a paper entitled "Remarks on 'Hog Flu'" was presented (Dorset et al. 1922). This group acknowledged that the name 'flu' given by Koen seemed to be appropriate, but suggested that the name 'hog flu' should be used to avoid confusion with human influenza. We were confronted with this issue more than 90 years later during the 2009 H1N1 pandemic where misunderstandings associated with the name "swine flu" led to considerable losses to the swine industry. Despite consistent and clear evidence to the contrary, people became fearful of catching pandemic influenza by eating pork (Joint FAO/WHO/OIE 2009).

Regardless of the name, SI continued to be a problem in swine herds throughout the twentieth century. In 1927, McBryde reported on the nature of the disease including the sudden onset with the entire herd having the disease within a couple of days. Clinical signs noted (similar to today) included fever, loss of appetite, lethargy, "thumpy" respiration, cough, loss of condition, and low mortality followed by quick recovery (McBryde 1927). Koen (1919) and other authors throughout the 1920s consistently reported that outbreaks of SI were associated with high morbidity and low mortality, which is also consistent with the disease as observed today. A major advance during that time was the ability to differentially diagnose hog cholera and swine influenza. As related to Dr. Easterday by the late S.H. McNutt, former mentor and colleague, a crude field method of differentiation was based simply on mortality patterns—"We finally determined how to differentiate between hog cholera and swine influenza—if most of the affected animals died they had cholera and if most of them lived they had flu."

Historically, influenza in swine was a disease of late autumn and early winter. However, with contemporary swine production methods including extensive climate control and animal confinement, the disease now can appear throughout the year rather than only seasonally. In the early years of the disease the farmers were very concerned and called their veterinarians on a regular basis for help. However, after learning that there was no specific treatment and that the animals would typically recover with little or no death loss, this has become a disease that is managed in many cases without direct veterinary medical input.

Even in the 1920s it was not uncommon to meet farmers and veterinarians who believed that they had contracted the disease from affected swine. In 1928, McBryde et al. reported on the transmission of SI by infectious materials from an affected animal to a normal animal and by placing normal animals with affected animals (McBryde et al. 1928). They concluded that the infectious agent was in nasal, tracheal, and bronchial fluids and that the infectious agent was not in the blood. They speculated that because of the sudden onset and that a large number of animals are simultaneously affected, some organism already present in the respiratory tract was the causative agent and was activated when the resistance of the animals was lowered by the harsh environment. They believed that there was little if any immunity, because "…it seems to be quite well established that the same herd may suffer from two or even three attacks of flu in one season." In 1931, Shope proved that swine influenza was caused by a virus and that he could reproduce SI under strict experimental conditions by inoculating both filtered and unfiltered material from affected pigs into the respiratory tract of normal pigs (Shope 1964). Subsequently, Shope would describe SI as a "…disease of complex etiology, being caused by infection with the bacterium *H. influenzae suis* and the swine influenza virus acting in concert." (Although we now understand that the virus alone is sufficient to cause the disease). Subsequently, in 1934, Andrewes et al. confirmed that human influenza was also associated with a virus (Andrewes et al. 1934).

Shope's work in the 1930s and 1940s was instrumental in our understanding of influenza. He demonstrated that SI could infect mice (Shope 1935) and ferrets (Shope 1934) and that humans had neutralizing antibodies for SI (Shope 1936), suggesting that the human influenza virus was antigenically similar to the swine influenza virus. Particularly intriguing was his hypothesis on how SI was maintained in nature as a seasonal disease. While the *H. influenzae suis* bacterium could be found throughout the year, with the methods available at the time the virus was not demonstrated during the 9-month interepizootic period. He hypothesized that the virus was harbored and transmitted in swine lungworms (*Metastrongylus*) and performed experiments which he believed demonstrated that the virus remained in an occult or masked form in the lungworms and was provoked to infectivity by an adverse meterological condition. However, the virus could not be detected by direct means in the lungworm larvae, in the earthworm intermediate host, or in the adult lungworm in the definitive host (Shope 1941a, b, 1943a, b, 1955). Nonetheless, that hypothesis served to stimulate many to further investigate SI. It is now recognized that the virus circulates throughout the year and that there is no conclusive evidence for a complex invertebrate host system to maintain the disease.

The unresolved question of the relationship of swine and human influenza and their respective viruses fueled speculation on the role of animals in human influenza. When the human pandemic of 1957 first began to spread in Asian countries, the World Health Organization (WHO) decided to perform an animal serum survey to better understand the role of animals in the epidemiology of influenza (Kaplan and Payne 1959). This decision was based on reports from China of epizootics of influenza-like disease in swine in areas severely affected by the human disease. Veterinary medical service agencies in 33 countries throughout the world participated in the survey by collecting serum from swine and horses. As requested by the WHO, the testing laboratories performed complement–fixation and hemagglutination-inhibition tests with provided reagents. These studies demonstrated that the Asian ("A2" at the time and in the old influenza nomenclature) strain caused naturally occurring, but often inapparent, infection in horses and swine. And following recognition of infection in pigs with the A-swine strain, long known in the USA, in Germany, and Czechoslovakia (Kaplan and Payne 1959), Kaplan and Payne (1959) stated that "For the clarification of the natural history of influenza, so urgently needed, this problem of animal influenza can no longer be neglected. Investigations along the lines indicated above will certainly add much to our knowledge of influenza epidemiology, and the World Health Organization hopes to stimulate and coordinate such studies in the future".

The WHO Expert Committee on Respiratory Virus Diseases met in 1958 and devoted most of the five-day meeting to influenza (Expert Committee on Respiratory Virus Diseases 1959). They considered the relationship between Asian influenza virus and viruses infecting animals, and the role, in general, of animals in the epidemiology of human influenza. During the course of that meeting, a combined session with the Joint WHO/FAO Expert Committee on Zoonoses was held. As a result of the considerable discussions, both independently and together, both agencies concluded that further investigations on the relationship between human and animal influenza viruses were needed. In 1960, Steele proposed the possibility of an animal reservoir for influenza A viruses may exist in nature (Steele 1961) and greater emphasis thereafter was placed on surveillance in animals.

Dr. Martin M. Kaplan served as the first Chief of Veterinary Public Health for the WHO from 1949 to 1969 and Director of the Office of Science and Technology (eventually to be called the Office of Research Promotion and Development) from 1969 until his retirement in 1976. He provided strong international leadership in influenza surveillance in animals and organized a worldwide survey in 1957. Kaplan convened the first WHO Informal Meeting on the Coordinated Study of Animal Influenza in Prague in January 1963. Such meetings were expanded and continued into the early 1980s. Because of the continued prominence of questions about the role of swine and swine influenza virus in human influenza, the participants in that first meeting agreed that the major efforts should be directed toward epidemiological studies in swine and to standardizing test reagents and technical procedures. The technical procedures focused on minimizing the possibility of laboratory contaminations in virus isolation and providing standard reagents and procedures for serological studies to reduce the problems of non-specific inhibitors in animal sera. Collaborators were expected to provide data on a regular basis to Dr. Kaplan and he, in turn, summarized the material and provided it to all of the collaborators.

The next meeting occurred in 1964 and the major topics of discussion at that meeting were swine influenza and equine influenza with other areas being avian influenza and the surveillance in mammals other than swine and horses. Antigenic analysis of influenza viruses was becoming a larger part of the collaboration, but the group agreed that for a variety of reasons, including the pitfalls of serological testing, there needed to be greater emphasis on the isolation of viruses.

Although significant research with swine and swine influenza continued, an increasing number of isolations of influenza viruses from a variety of avian species led to expanded collaborations on the epidemiology of influenza in birds. WHO was able to provide support for some part of the research activities at most of the collaborating laboratories (Kaplan 1969); however, it should be recognized that a major basis for the success of these collaborations was the collegiality and trust that developed among the participants. This included generous sharing of data as it was developed and opportunities to visit the various participating laboratories and institutions. The collaborative efforts resulted in many joint publications involving participants from two or more laboratories.

By 1970 considerable evidence had accumulated, based on serological studies, that people whose occupations brought them in contact with swine became infected with the swine influenza virus. In 1974, the swine influenza virus was isolated from the lung of a boy who had died with Hodgkin's disease (Easterday 1986). He had been in contact with swine five days before he died and the swine had antibody to the virus. An acute respiratory disease in an 8-year-old boy in Wisconsin in 1975 was attributed to infection with swine influenza virus based on serological studies of serum from the boy and the swine with which he had been in contact (Easterday 1986). Then came the Fort Dix, New Jersey swine influenza virus "episode" beginning in January 1976 in which recruits fell ill with respiratory illness ultimately shown to be due to infection with a swine influenza virus (Gaydos et al. 2006). All speculation about the transmission of the virus from swine to human beings came to an end when the virus was isolated from swine and their caretaker on a Wisconsin farm in the autumn of 1976 (Easterday 1986), and a recent comprehensive review article by Myers et al. documents multiple cases of human infection with SI viruses since that time (Myers et al. 2007).

Much remains to be learned about SI as a disease of swine. It continues to be a problem in swine production and the high morbidity rate with acute illness in the swine is estimated to result in a delay of as much as two weeks in the affected animals reaching market weight. That delay results in increased production costs. Thus, efforts have been aimed toward prevention of infection. Swine influenza vaccines have been used commercially in swine herds in the US since the early 1990s. During the past several years, with the emergence of multiple subtypes and genotypes not previously circulating in swine in the US, manufacturers have updated the composition of the vaccines to include the new strains. Due to documented cases of reverse zoonotic transmission of the 2009 pH1N1 virus to pigs, this included development of a USDA conditionally licensed pandemic H1N1 strain vaccine for pigs. The choice to vaccinate in the US remains a decision between producers and their herd veterinarians. In addition to the commercially available vaccines in the US, it has been estimated that more than 50% of the vaccines that are used are autogenous products that are custom-created for individual swine production units. Swine are also vaccinated against SI in many other parts of the world, though the specific virus strains in vaccines and role of vaccination vary with the swine production management systems.

Prior to the late 1990s, the 1918-derived H1N1 virus was the predominate agent of SI circulating among swine in North America. However, since the late 1990s, we have seen the emergence of novel strains of influenza viruses in swine in North America, including most prominently, triple reassortant H3N2, H1N2, and H1N1 viruses (reviewed in Olsen 2002; Vincent et al. 2008), as well as wholly avian H4N6, H3N3, and H1N1 swine isolates (Karasin et al. 2000, 2004). These events, as well as the recognition that the 2009 pandemic H1N1 virus had its genetic origins in viruses of swine influenza origins, should serve as a wake-up call for the world's animal health and public health communities. One must remain vigilant to the constant emergence of new influenza viruses in animals and surveillance of influenza in animals and humans would be well served by an integrated system and "one health" approach.

References

Andrewes CH, Laidlaw PP, Smith W (1934) The susceptibility of mice to the viruses of human and swine influenza. Lancet 224:859–862

Dorset M, McBryde CN, Niles WB (1922) Remarks on "hog flu". J Am Vet Med Assoc 62:162–171

Easterday BC (1986) Swine influenza. In: Leman AD et al (eds.) Diseases of swine, 6th edn. Iowa State University Press, Ames, pp 244–255

Expert Committee on Respiratory Virus Diseases (1959) First report. World Health Organization Technical Report Series, number 170, pp 60

Gaydos JC, Top FH, Hodder RA, Russell PK (2006) Swine influenza a outbreak, Fort Dix, New Jersey, 1976. Emerg Infect Dis 12:23–28

Joint FAO/WHO/OIE (2009) Joint FAO/WHO/OIE Statement on influenza A(H1N1) and the safety of pork. http://www.who.int/mediacentre/news/statements/2009/h1n1_20090430/en/index.html
Kaplan MM, Payne AMM (1959) Serological survey in animals for type A influenza in relation to the 1957 pandemic. Bull World Health Organ 20:465–488
Kaplan MM (1969) Relationships between animal and human influenza. Bull World Health Organ 41:485–486
Karasin AI, Brown IH, Carman S, Olsen CW (2000) Isolation and characterization of H4N6 avian influenza viruses from pigs with pneumonia in Canada. J Virol 74:9322–9327
Karasin AI, West K, Carman S, Olsen CW (2004) Characterization of avian H3N3 and H1N1 influenza a viruses isolated from pigs in Canada. J Clin Microbiol 42:4349–4354
Koen JS (1919) A practical method for field diagnosis of swine diseases. Am J Vet Med 14:468–470
McBryde CN (1927) Some observations on "hog flu" and its seasonal prevalence in Iowa. J Am Vet Med Assoc 71:368–377
McBryde CN, Niles WB, Moskey HE (1928) Investigations on the transmission and etiology of hog flu. J Am Vet Med Assoc 73:331–346
Murray C (1921) What is "hog flu?" Wallaces' Farmer 46:371
Myers KP, Olsen CW, Gray GC (2007) Cases of swine influenza in humans a review of the literature. Clin Infect Dis 44:1084–1088
Olsen CW (2002) The emergence of novel swine influenza viruses in North America. Virus Res 85:199–210
Shope RE (1934) The infection of ferrets with swine influenza virus. J Exp Med 60:49–61
Shope RE (1935) The infection of mice with swine influenza virus. J Exp Med 62:561–572
Shope RE (1936) The incidence of neutralizing antibodies for swine influenza virus in the sera of human beings of different ages. J Exp Med 63:669–684
Shope RE (1941a) The swine lungworm as a reservoir and intermediate host for swine influenza virus: I. The presence of swine influenza virus in healthy and susceptible pigs. J Exp Med 74:41–47
Shope RE (1941b) The swine lungworm as a reservoir and intermediate host for swine influenza virus: II. The transmission of swine influenza virus by the swine lungworm. J Exp Med 74:49–68
Shope RE (1943a) The swine lungworm as a reservoir and intermediate host for swine influenza virus: III. Factors influencing the transmission of the virus and the provocation of influenza. J Exp Med 77:111–126
Shope RE (1943b) The swine lungworm as a reservoir and intermediate host for swine influenza virus: IV. The demonstration of masked swine influenza virus in lungworm larvae and swine under natural conditions. J Exp Med 77:127–138
Shope RE (1955) The swine lungworm as a reservoir and intermediate host for swine influenza virus: V. Provocation of swine influenza by exposure to prepared swine to adverse weather. J Exp Med 102:567–572
Shope RE (1964) Swine influenza. In: Dunne HW (ed) Diseases of swine, 2nd edn. Iowa State University Press, Ames, pp 109–126
Steele JH (1961) Animal influenza. Am Rev Respir Dis 83:41–46
Vincent AL, Ma W, Lager KM, Janke BH, Richt JA (2008) Swine influenza viruses a North American perspective. Adv Virus Res 72:127–154

Genetics, Evolution, and the Zoonotic Capacity of European Swine Influenza Viruses

Roland Zell, Christoph Scholtissek and Stephan Ludwig

Abstract The European swine influenza virus lineage differs genetically from the classical swine influenza viruses and the triple reassortants found in North America and Asia. The avian-like swine H1N1 viruses emerged in 1979 after an avian-to-swine transmission and spread to all major European pig-producing countries. Reassortment of these viruses with seasonal H3N2 viruses led to human-like swine H3N2 viruses which appeared in 1984. Finally, human-like swine H1N2 viruses emerged in 1994. These are triple reassortants comprising genes of avian-like H1N1, seasonal H1N1, and seasonal H3N2 viruses. All three subtypes established persistent infection chains and became prevalent in the European pig population. They successively replaced the circulating classical swine H1N1 viruses of that time and gave rise to a number of reassortant viruses including the pandemic (H1N1) 2009 virus. All three European lineages have the capacity to infect humans but zoonotic infections are benign.

R. Zell
Department of Virology and Antiviral Therapy, Jena University Hospital, Friedrich Schiller University, D-07740 Jena, Germany
e-mail: roland.zell@med.uni-jena.de

C. Scholtissek
Institute of Virology, Justus- Liebig-University, 35392 Giessen, Germany

S. Ludwig (✉)
Institute of Molecular Virology, Centre of Molecular Biology of Inflammation, Westfälische Wilhelms University, D-48161 Münster, Germany
e-mail: ludwigs@uni-muenster.de

C. Scholtissek
Present address: Waldstr. 53, D-35440 Linden, Germany

Current Topics in Microbiology and Immunology (2013) 370: 29–55

DOI: 10.1007/82_2012_267

Published Online: 26 September 2012

Contents

1 Introduction

Swine influenza was first recognized as a disease of pigs during the great pandemic in autumn 1918. At that time, John S. Koen, who worked as a hog cholera inspector for the U.S. Bureau of Animal Industry in Fort Dodge, Iowa, observed a striking similarity between the clinical presentation of diseased humans and pigs: "*Last fall and winter we were confronted with a new condition, if not a new disease. I believe I have as much to support this diagnosis in pigs as the physicians have to support a similar diagnosis in man. The similarity of the epidemic among people and the epidemic in pigs was so close, the reports so frequent, that an outbreak in the family would be followed immediately by an outbreak among the hogs and* vice versa, *as to present a most striking coincidence if not suggesting a close relation between the two conditions. It looked like "flu", it presented the identical symptoms of "flu", it terminated like "flu", and until proved it was not "flu", I shall stand by that diagnosis.*" (Koen 1919). The etiologic agent of "flu", influenza A virus, was first isolated by Richard E. Shope (Shope 1931a).

Influenza viruses are members of the family *Orthomyxoviridae* which comprises five genera: *Influenza virus A*, *B*, and *C*, *Thogotovirus*, and *Isavirus* (Kawaoka et al. 2005). Each influenza virus genus includes one species (also designated as *influenza A virus*, *influenza B virus*, and *influenza C virus*; abbreviated FLUAV, FLUBV, FLUCV).

Influenza A viruses are enveloped negative-stranded RNA viruses. The RNA genome is segmented (Duesberg 1968) and associated with the viral nucleoprotein

(NP) and the viral polymerase complex. The eight RNA segments vary in their sizes (ranging from 890 to 2,341 nucleotides) and encode 11 proteins. Expression of viral genes occurs after transcription of genomic RNA with the help of the viral RNA-dependent RNA polymerase.

New isolates of influenza A virus predominantly have a filamentous structure with a diameter of approximately 80–120 nm and a length ranging from 2 to 200 μm (Chu et al. 1949). After adaptation to cell culture, the virion tends to have a spherical or pleomorph appearance. The viral envelope is composed of the cell membrane lipids but the majority of surface proteins are provided by the viral hemagglutinin (HA) and neuraminidase (NA). A third viral membrane protein is the M2 proton channel. The inner surface of the envelope is coated with the matrix protein (M1). The virion contains eight nucleocapsids; these are complexes of RNA and viral protein. Electron micrographs show helical rod-like structures with a terminal loop. The width ranges from 10 to 15 nm and the length from 30 to 120 nm. They were interpreted to represent backfolded and twisted ribonucleoproteins (Compans et al. 1972). The nucleocapsids are associated with the envelope by matrix proteins (Noda et al. 2006).

2 Influenza Virus Ecology

Influenza A viruses have a broad host range (Webster et al. 1992). The main reservoir hosts are aquatic birds of the orders *Anseriformes* (geese and ducks) and *Charadriiformes* (waders and gulls), but numerous other bird species may also be infected (Munster et al. 2007). Reassortment of 16 HA and nine NA types allows the formation of maximal 144 HA/NA combinations of which more than 110 types have been already isolated from birds. Whether all theoretical combinations exist in nature is unknown. In mammals, stable infection chains are observed only for certain subtypes (Table 1). Important mammalian host species include: humans, pigs, and horses. Dogs, domestic cats, and felid carnivores (tiger, leopard) as well as several mustelid carnivores (ferret, stone marten, mink), marine mammals (whales, seals), the camel, the muskrat, civet, racoon dog, pika, and giant anteater were described as accidental hosts without establishment of stable infection chains.

Influenza virus ecology is strongly influenced by virus adaptation to its host. One major determinant of the host range is the receptor molecule on the surface of the host cell. Influenza A virus binds to sialic acid (N-acetylneuraminic acid) which is linked by an α-glycosidic bond to the terminal galactose residues of carbohydrate chains of glycoproteins and glycolipids (Rogers and Paulsen 1983). Both species and tissue-specific expression of receptor molecules determine host range and tropism of influenza A viruses (Ito et al. 1998; Ito and Kawaoka 2000). Whereas avian influenza viruses bind to α-2,3-linked sialic acid, seasonal influenza virus strains of humans recognize α-2,6-linked sialic acid. Airway epithelia of the upper respiratory tract of pigs express both receptors. Thus, pigs are susceptible to

Table 1 Mammalian host species of influenza A virus

Host	Stable infection chain	Incidental zoonotic infection or reassorted isolate	Extinct
Human	H1N1, H3N2, pandemic (H1N1) 2009 virus	H1N2, H5N1, H7N1, H7N2, H7N3, H7N7, H9N2	H2N2
Pig	H1N1, H1N2, H3N2	H1N7, H2N3, H3N1, H3N3, H4N6, H5N1, H5N2, H9N2, pandemic (H1N1) 2009 virus	
Horse	H3N8	H1N8, H3N3	H7N7
Dog		H3N8, H5N1	
Racoon dog (*Nyctereutes procyonoides*)		H5N1	
Mink		H3N2, H10N4	
Stone marten		H5N1	
Ferret		H1N1	
Seal (*Phoca vitulina*)		H3N3, H4N5, H7N7	
Whale		H1N3, H13N2, H13N9	
Camel		H1N1	
Giant anteater		H1N1	
Tiger, leopard		H5N1	
Domestic cat		H5N1, pandemic (H1N1) 2009 virus	
Civet		H5N1	
Pika (*Ochotona spec.*)		H5N1	
Muskrat (*Ondatra zibethicus*)		H4N6	

influenza viruses which are adapted either to birds or to humans and can serve as intermediate hosts after trans-species infections (Ito et al. 1998). Due to their receptor configuration, pigs were considered as mixing vessels for the reassortment of human and avian influenza viruses (Scholtissek et al. 1985). Other host determining factors are the nucleoprotein (Scholtissek 1990) and polymerase subunit PB2. Amino acid position 627 of PB2 was shown to be critical for virus replication (Subbarao et al. 1993). The tissue-specific expression of host proteases, however, contributes to virulence or pathogenicity but not to the host range (Webster et al. 1992; Steinhauer 1999). Recent genome-wide RNAi screening studies revealed the involvement of hundreds of host factors that are required for efficient influenza virus replication (König et al. 2010; Karlas et al. 2010). It remains to be elucidated which of these factors establishes the host range.

3 Genetic Drift and Reassortment: Two Mechanisms for the Generation of Genetic Variability of Influenza viruses

A typical property of influenza A viruses is their great variability which is mainly caused by two mechanisms. Genetic drift is the continuous accumulation of nucleotide substitutions over time. The substitution rate of influenza viruses ranges from 10^{-5} to 10^{-6} substitutions/site/replication cycle depending on the experimental setup (e.g., Stech et al. 1999; Nobusawa and Sato 2006; Parvin et al. 1986). According to the genome size of appr. 13,600 nucleotides, between 1 and 10 % of the progeny virus has substitutions. Older estimations determined even higher substitution rates (e.g., Yewdell et al. 1979; Holland et al. 1982). The most base substitutions are neutral, this means they do not cause changes of the amino acid sequence or—if so—a substitution does not seem to influence the fitness of the progeny virus. The reason is that the majority of amino acid residues of influenza virus proteins are negatively selected (purifying selection). Substitutions of such amino acids would decrease the viral fitness and are only endured as long as certain selection pressures act on the virus. A host change could induce such a selection pressure. Only very few sites are positively selected. A positive selection increases the heterogeneity of the gene pool, it is also designated as ‘diversifying selection’. One example of positive selection is the gradual changing of the antigenic sites of the hemagglutinin known as ‘antigenic drift’. Substitutions that result in immune escape variants have an increased probability to infect hosts with preimmunity. Eighteen codons of the hemagglutinating HA1 domain were identified to be positively selected (Bush et al. 1999a, b). One driver of antigenic drift is the receptor binding avidity of the viral hemagglutinin (Hensley et al. 2009). Though antigen drift of the hemagglutinin may be striking and can be investigated by serological means, substitutions of all influenza virus genes occur with the same frequency. Accordingly, the whole genome is subjected to a genetic drift rather than antigenic drift of the HA gene only. Due to the preponderance of either positive or negative selection acting on each gene, the relation of synonymous and non-synonymous substitutions of the eight gene segments differs.

The second important mechanism of influenza virus variability is reassortment or the exchange of one or more gene segments. Reassortments are of biological importance as they lead to novel combinations of genome segments which have been evolved by negative or positive selection. This mechanism greatly enhances evolutionary rates and accounts for rapid viral adaptation to changing environmental conditions. Reassortments occur naturally or can be induced experimentally (Kilbourne 1968). They are accomplished by a segmented virus genome and by double or multiple infections of a host with virus strains of different subtypes or genetic lineages. Reassortment events leading to exchanges of HA and NA genes are of special importance as they can lead to an antigen shift. Shift variants exhibit major differences of antigenic epitopes and less cross-reactivity with pre-existing antibodies of a host. Circulation of two or more subtypes within a population at the same time can lead to reassortments which are not associated with an antigen shift.

In addition, reassortments may occur after incidental zoonotic infections. Such events introduce genes into a virus population that are adapted to other species. Beside the HA and NA genes, other gene segments can also reassort but were in the dark for long time due to a lack of sequence data. Such reassortants are serologically inconspicuous.

4 The Concept of Genetic Influenza A Virus Lineages

Genetic drift did not only contribute to the evolution of the known HA and NA subtypes, but led to the formation of distinct genetic lineages of all genome segments (Webster et al. 1992). The genetic configuration of an influenza virus is defined by its genotype (Lu et al. 2007), which describes a virus with greater accuracy than subtyping by the HA and NA types only. Precise genotyping requires complete genome sequences but greatly enhanced our understanding of influenza virus ecology and evolution. Whereas sequence comparisons of different HA types yield nucleotide identities of roughly 56 % on average, sequences of a given lineage have nucleotide identities greater than 90 %. Two factors determine the evolution of genetic lineages: host species barriers (Kuiken et al. 2006) and geographic isolation. Starting with the pandemic of 1918, two stable infection chains of H1N1 were established in humans and pigs which lead to new, distinct genetic lineages: the seasonal H1N1 of humans and the classical swine H1N1 lineage (Fig. 1). Such lineages can be demonstrated for all eight genome segments. There are distinct lineages for birds, humans, pigs, and horses. It appears that some lineages became extinct but the significance of this observation is yet unclear due to a lack of sufficient sequence data. Viruses of a lineage are adapted to their host species. Trans-species infections occasionally occur, but virus replication is less efficient and infection chains may disrupt after a few generations. Stable infection chains will establish rarely. Besides the pandemic virus of 1918, the 'avian-like' swine influenza viruses in Europe are another example for a successful establishment of a stable infection chain. However, after some 30 years of circulation the latter viruses have not yet accumulated sufficient substitutions to establish distinct genetic lineages for each of their genes. Their H1 HA gene, for example, is presently considered as a sublineage or clade of lineage 1C (Fig. 1). The genes of the present pandemic (H1N1) 2009 virus have the potential to lead to new genetic lineages. Among the 16 HA types at least 69 genetic lineages were described; the nine NA types comprise altogether 46 lineages and each of the internal segment has 7–11 lineages (Lu et al. 2007).

In addition to host species barriers, geographic isolation can induce the development of genetic lineages. As a result of different flyways of migratory birds, American and Eurasian lineages of influenza virus genes evolved (Olsen et al. 2006). As the evolution of such lineages is promoted by isolation rather than host-specific barriers, trans-species infections are not uncommon (Krauss et al. 2007; Wallensten et al. 2005; zu Dohna et al. 2009). They occur frequently in

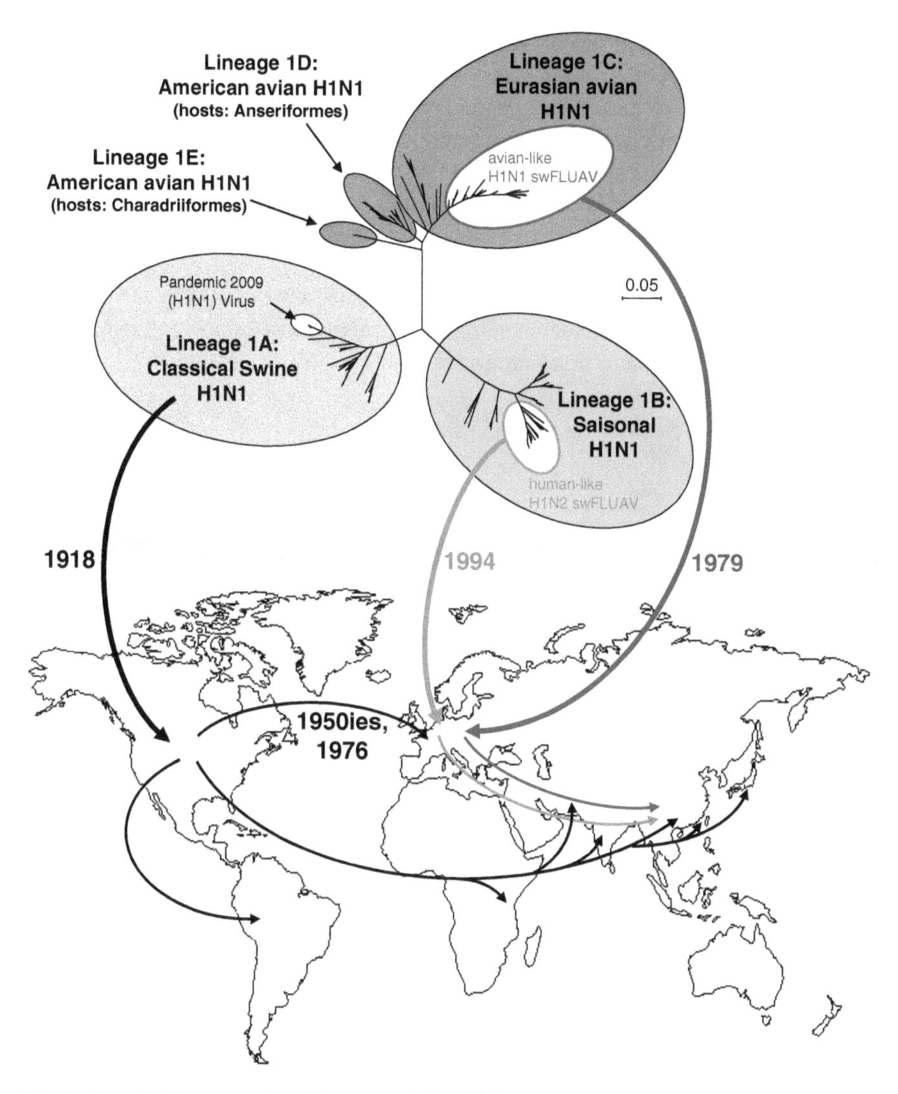

Fig. 1 Genetic lineages and sublineages of the HAH1 gene and their geographic distribution. Adaptation to host species and geographic isolation lead to the evolution of five HAH1 lineages (*top*). Genetic data suggest that the seasonal H1N1 viruses of humans emerged around 1918 from an American avian ancestor, either as a whole or as a reassortant (Anhlan et al. 2011), and spread worldwide (pandemic of 1918, not indicated). Classical swine H1N1 viruses also emerged around that time in the USA, probably after zoonotic infections of pigs. In several waves classical swine viruses were translocated to South America, numerous Asian countries, East Africa, and Europe. In Europe, they vanished after appearance of the avian-like H1N1 swine viruses in Belgium and Germany in 1979. Human-like H1N2 swine influenza viruses emerged in 1994 in the UK and spread to the European continent. After 2000, avian-like H1N1 and human-like H1N2 arrived in Asia where they co-circulate with the classical swine strains. Numerous reassortants indicate a dynamic influenza activity in Asia

overlapping breeding grounds, for example around the Arctic Beringia Sea (Wahlgren et al. 2008) or after translocation of infected birds (Makarova et al. 1999). Emergence of new lineages or subtypes may lead to the extinction of previously circulating types, a phenomenon that was repeatedly observed but cannot be sufficiently explained yet. For example, seasonal H1N1 viruses were superseded by pandemic H2N2 viruses in 1957. Likewise, circulating classical swine H1N1 viruses were replaced by avian-like swine H1N1 viruses after 1979 in Europe.

Despite the generation of thousands of sequence entries in the GenBank in recent years, our present understanding of the dynamics of the influenza virus epizootiology of birds and non-human mammals is still fragmentary.

5 The Disease

Swine influenza was originally described as a disease of autumn and early winter which occurred in annual epizootics (Shope 1931b). In many regions with dense pig populations the disease became enzootic and nowadays infections occur all year-round. The main symptoms of swine influenza are sudden onset of the disease, fever, anorexia, coughing, nasal discharge, sneezing, dyspnoea, exhaustion, and apathy. In general, infections with the virus cause a mild disease with a benign outcome. The morbidity within an affected herd is high (up to 100 %); the mortality is low but depends on the virus strain and other factors such as mixed infections. Usually the disease lasts 2–6 days and in most cases animals completely recover. Affected pigs develop an acute bronchitis with swollen mucosa, abundant mucus, hyperemia, and enlarged local lymph nodes. Inflammation surrounds bronchi and bronchioles. Sometimes secondary bacterial lobular pneumonia exacerbates the disease and may lead to death. Koen (1919) estimated influenza-associated mortality with "*1 per cent, at any rate less than 2 per cent*"; Shope with 1–4 % (Koen 1919; Shope 1931b).

In Europe, swine influenza is caused by three virus subtypes that are genetically distinct from the classical swine H1N1 viruses. In regions with enzootic persistence, the clinical signs are less marked and the virus circulates throughout the year. The avian-like swine H1N1 viruses generally induce less severe symptoms than human-like swine H3N2 viruses and natural H1N1 infections are sometimes unrecognized. The European H1N2 strains differ regarding their virulence. For example, the German strains which appeared in 2000 are more virulent than the Belgian strains. Changes in epizootiology may have several reasons. One reason is that swine husbandry practices changed in the past decades. A short fattening period of only 6 months leads to a rapid turnover of the swine population in a herd which requires purchase of piglets from various suppliers. This increases the chance of a virus to infect naive pigs and accomplishes gathering of influenza viruses from several distinct sources. Less marked clinical signs of enzootic viruses, varying levels of maternal antibodies, and preexisting immunities of older pigs constitute selection pressures which are at the molecular level yet undefined and prepare the ground for the reassortment of novel virus combinations.

6 European Swine Influenza Viruses

There is evidence that influenza viruses have been introduced several times into European pigs. Stable infection chains, however, were first established in the 1970s. Prior to this, seasonal and classical swine influenza viruses have been detected in serological studies and occasional virus isolations. In this review we distinguish between zoonotic viruses of human and avian origin isolated from pigs on the one hand and the "*human-like*" and "*avian-like*" viruses on the other hand. Whereas the former viruses occur occasionally and become extinct after a few replication cycles, the latter viruses have defined genetic settings which developed distinct clades in phylogenetic trees. Such clades indicate stable infection chains over many virus generations. Moreover, these viruses exhibit evolutionary changes as a result of genetic drift and/or reassortment events which will be reviewed here. Previous authors (e.g., Brown 2000; Kuntz-Simon and Madec 2009) used the terms "*human-like*" and "*avian-like*" primarily to indicate the previous host.

6.1 Early Descriptions of Swine Influenza in Europe

Several early reports describe the influenza of pigs in Europe. Soon after the first isolation of swine influenza virus by Richard E. Shope, K. Köbe published the isolation of the etiologic agent of "enzootic pneumonia" of piglets, a condition that he named "Ferkelgrippe" (Köbe 1933). Köbe had observed that the histopathological lesions of the lungs were like "American swine influenza", but "Ferkelgrippe" in Germany showed enzootic rather than epizootic transmission and occurred in piglets younger than those of Shope's experiments. In analogy to Shope's work on swine influenza, Köbe believed that "Ferkelgrippe" was the result of a mixed infection; experimental pneumonia was induced only after coinfection of piglets with a filtrable virus that he had isolated from dispersed lung tissue of affected pigs and a bacterium that he designated *Bacterium influenzae suis*. The virus alone induced only mild symptoms and was distinct from the classical swine fever virus (hog cholera virus). It is unknown whether Köbe tried to propagate the virus in ferrets or mice. Unfortunately, the virus became lost in the past decades. Köbe's mentor Otto Waldmann confirmed the findings of his associate and commented that the observed differences in age incidence could be due to different husbandry practices in Germany and the USA (Waldmann 1933). Shortly thereafter, Gerhard Elkeles in Berlin, Germany, infected 2–6-week-old piglets with human influenza virus and could induce a mild flu-like disease in pigs; experimental coinfection of piglets with the human virus and either human or porcine strains of *Haemophilus influenzae* resulted in a more severe disease (Elkeles 1934). These were the first experiments demonstrating the susceptibility of pigs to human influenza virus strains. Later they were confirmed by Shope and Francis (1936), but these authors used older pigs (6–14 weeks). With regard to the findings of Köbe and Elkeles, they discussed a different natural susceptibility of European and American pigs. Further work was published by Lamont (1938) and Blakemore and Gledhill (1941) who described outbreaks of swine influenza in Northern Ireland and England. Interestingly, Blakemore and Gledhill (1941) observed one outbreak on an Essex farm with cases of chronic disease (for 8 weeks), and—like O. Waldmann—concluded that husbandry conditions may have an influence on the course of the disease. Both Lamont and Blakemore handed over tissue specimens of several outbreaks to R. E. Glover in Cambridge who together with C. H. Andrewes succeeded to isolate three influenza virus strains after serial passages in ferrets and mice (Glover 1941). Serological characterization revealed that these isolates differed from Shope's swine influenza virus but resembled the human strains. Later, three of Glover's virus strains of that time were (partially) sequenced and were shown to cluster with A/WS/1933 and with A/Puerto Rico/8/1934 (Gorman et al. 1991; Neumeier and Meier-Ewert 1992; Neumeier et al. 1994; Yoshioka et al. 1994).

There is no hard evidence that classical swine influenza virus entered Europe in the 1930s or 1940s. European strains of classical swine influenza virus were first isolated in the 1950s in the former Czechoslovakia (Harnach et al. 1950). A serological survey conducted in 1957 by the W.H.O. revealed antibodies in pigs to classical swine H1N1 in Czechoslovakia and Germany (Kaplan and Payne 1959). After this episode, classical swine viruses disappeared in the 1960s in Europe, but were reintroduced in 1976 in Italy (Nardelli et al. 1978). These viruses spread to several European countries, including Belgium (Biront et al. 1980; Vandeputte et al. 1980), Germany (Sinnecker et al. 1983), France (Gourreau et al. 1980), England (Roberts et al. 1987), and Sweden (Martinsson et al. 1983). After the emergence of avian-like H1N1 swine viruses they disappeared again. The last European strain of classical swine H1N1 was isolated in 1993 in England soon after the first detection of avian-like H1N1 on the British Isles (Brown et al. 1997b).

6.2 *Stable Establishment of Influenza Viruses in European Pigs: Avian-Like Swine H1N1*

A distinct sublineage of European H1N1 swine influenza viruses emerged in January 1979 in Belgium (Pensaert et al. 1981). These viruses differed serologically from classical swine viruses but showed relationship to avian viruses. Some virulent strains induced clinical symptoms which were typical for swine influenza. In winter 1979/80, similar viruses appeared in Germany and France (Witte et al. 1981; Ottis et al. 1981; Gourreau et al. 1981). Retrospective serological analyses revealed that the majority of infections were asymptomatic. The molecular characterization revealed that all segments derived from an avian H1N1 influenza virus (Scholtissek et al. 1983). The phylogenetic comparison (Fig. 2a, b) demonstrates that the hemagglutinin of the swine viruses is most closely related to virus isolates from German ducks (A/duck/Bavaria/1/1977, A/duck/Bavaria/2/1977) which were the first avian H1N1 viruses detected in Europe (Ottis and Bachmann 1980). The hemagglutinin of the so-called "avian-like" swine H1N1 viruses shows a considerable cross-reaction with the classical swine H1N1; therefore, one of the commercially available vaccines (Gripovac™) for pigs includes A/New Jersey/8/1976 (H1N1). Further characterization revealed the avian origin of all segments (Schultz et al. 1991; Castrucci et al. 1993; Campitelli et al. 1997; Brown et al. 1997b). In very short time the avian-like H1N1 swine viruses established a stable infection chain and spread to all major swine-producing countries in Europe. They succeeded to replace the previous circulating classical swine strains. After 30 years of circulation, the avian-like swine H1N1 are endemic in the major pig-producing European countries. However, the seroprevalence varies considerably. In 2002/2003 it was highest in Belgium and Germany (80.8, 70.8 %); prevalence was lower in Italy and Spain; (46.4, 38.5 %) and low in the Czech Republic, Ireland, and Poland (>18 %) (Van Reeth et al. 2008). A more recent study conducted as a cross-sectional survey in Spain in 2008–2009 revealed a striking increase in the H1N1 seroprevalence (Simon-Grifé et al. 2010). Likewise, a German study indicates a similar annual variability suggesting fluctuations in the prevalence of swine influenza viruses over time (R. Dürrwald, personal communication).

6.3 *Emergence of Human-Like H3N2 in European Pigs*

The first "human-like" swine H3N2 virus emerged in Germany in 1982 (Schrader and Süss 2004). The HA and NA surface proteins of strain A/swine/Potsdam/35/1982 were derived from an A/Port Chalmers/1/1973-like seasonal H3N2 virus (Fig. 2c, d), whereas an avian-like H1N1 swine virus served as donor for the internal segments (M-segment: Schmidtke et al. 2006; Krumbholz et al. 2009; PB1-segment: Zell et al. 2007; PB2, PA, NP, NS segments: R. Zell unpublished). This virus disappeared soon. Another virus with a very similar genetic composition reemerged in 1984 and achieved to establish a persistent infection chain. The viruses were designated as "human-like swine H3N2" (Fig. 2c, d) due to their antigenic similarity to human H3N2. They spread rapidly in the European pig population. Epizootics were reported in Belgium (Haesebrouck et al. 1985; Haesebrouck and Pensaert 1988), France (Madec et al. 1984), and Germany (Zhang et al. 1989), Italy (Castrucci et al. 1993), the Netherlands (Loeffen et al. 1999), and Spain (Castro et al. 1988; Yus et al. 1992). The molecular analysis of these viruses revealed avian-like internal genes and human A/Port Chalmers/1/1973-like HA and NA genes (Campitelli et al. 1997; Marozin et al. 2002), but this

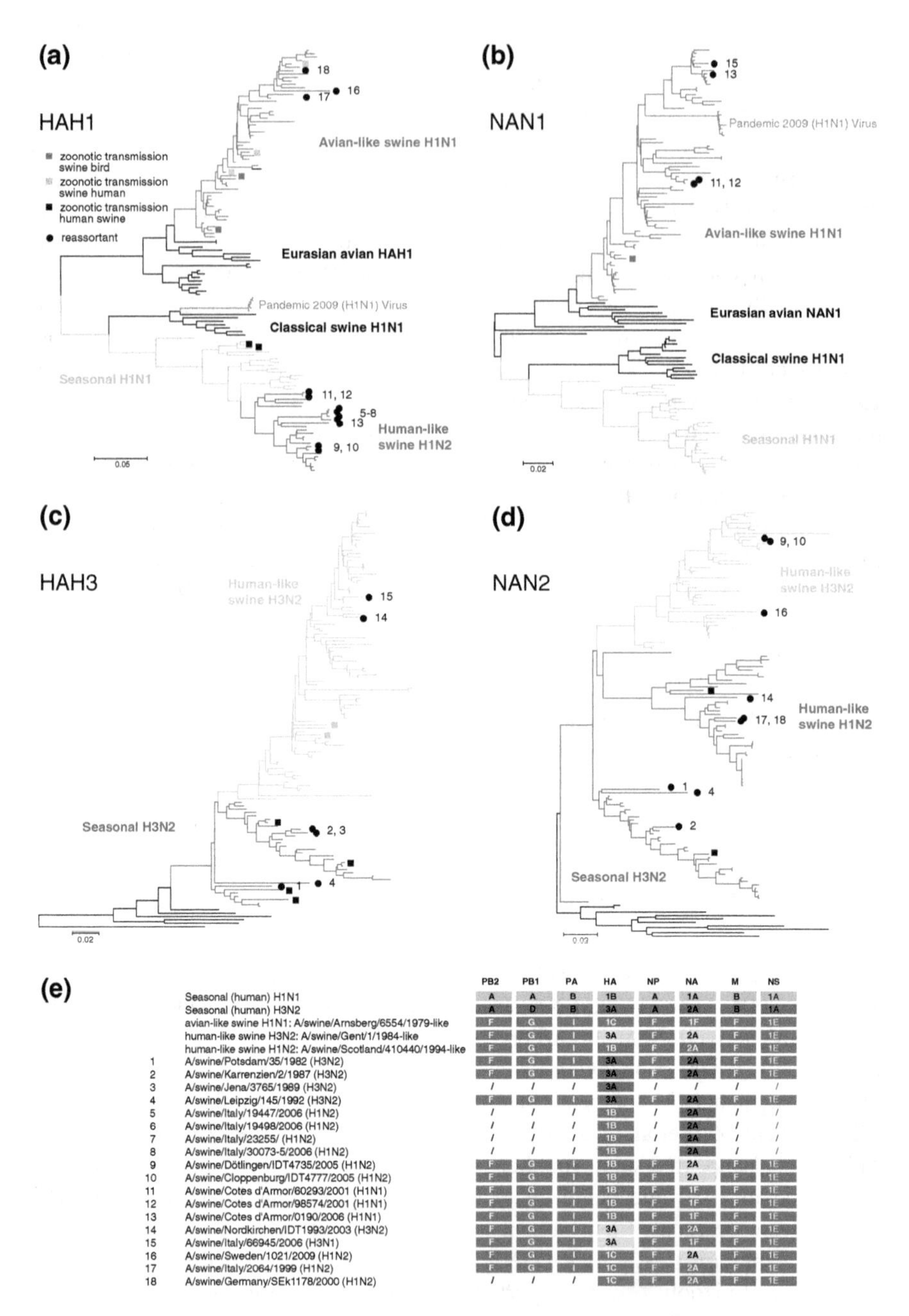

		PB2	PB1	PA	HA	NP	NA	M	NS
	Seasonal (human) H1N1	A	A	B	1B	A	1A	B	1A
	Seasonal (human) H3N2	A	D	B	3A	A	2A	B	1A
	avian-like swine H1N1: A/swine/Arnsberg/6554/1979-like	F	G	I	1C	F	1F	F	1E
	human-like swine H3N2: A/swine/Gent/1/1984-like	F	G	I	3A	F	2A	F	1E
	human-like swine H1N2: A/swine/Scotland/410440/1994-like	F	G	I	1B	F	2A	F	1E
1	A/swine/Potsdam/35/1982 (H3N2)	F	G	I	3A	F	2A	F	1E
2	A/swine/Karrenzien/2/1987 (H3N2)	F	G	I	3A	F	2A	F	1E
3	A/swine/Jena/3765/1989 (H3N2)	/	/	/	3A	/	/	/	/
4	A/swine/Leipzig/145/1992 (H3N2)	F	G	I	3A	F	2A	F	1E
5	A/swine/Italy/19447/2006 (H1N2)	/	/	/	1B	/	2A	/	/
6	A/swine/Italy/19498/2006 (H1N2)	/	/	/	1B	/	2A	/	/
7	A/swine/Italy/23255/ (H1N2)	/	/	/	1B	/	2A	/	/
8	A/swine/Italy/30073-5/2006 (H1N2)	/	/	/	1B	/	2A	/	/
9	A/swine/Dötlingen/IDT4735/2005 (H1N2)	F	G	I	1B	F	2A	F	1E
10	A/swine/Cloppenburg/IDT4777/2005 (H1N2)	F	G	I	1B	F	2A	F	1E
11	A/swine/Cotes d'Armor/60293/2001 (H1N1)	F	G	I	1B	F	1F	F	1E
12	A/swine/Cotes d'Armor/98574/2001 (H1N1)	F	G	I	1B	F	1F	F	1E
13	A/swine/Cotes d'Armor/0190/2006 (H1N1)	F	G	I	1B	F	1F	F	1E
14	A/swine/Nordkirchen/IDT1993/2003 (H3N2)	F	G	I	3A	F	2A	F	1E
15	A/swine/Italy/66945/2006 (H3N1)	F	G	I	3A	F	1F	F	1E
16	A/swine/Sweden/1021/2009 (H1N2)	F	G	I	1C	F	2A	F	1E
17	A/swine/Italy/2064/1999 (H1N2)	F	G	I	1C	F	2A	F	1E
18	A/swine/Germany/SEk1178/2000 (H1N2)	/	/	/	1C	F	2A	F	1E

Fig. 2 Phylogenetic trees. 125 representative sequences were aligned and used to infer the evolutionary relationships using the neighbor-joining method. Phylogenetic analyses were conducted in MEGA4 (Tamura et al. 2007). Avian-like swine influenza virus sequences are indicated in *blue*, human-like H3N2 in *ochre*, human-like H1N2 in *red*, seasonal H1N1 in *green*, and seasonal H3N2 in *pink*. Strain designations were omitted for clarity. **a** HAH1 sequences, **b** NAN1 sequences, **c** HAH3 sequences, **d** NAN2 sequences, **e** genetic composition of seasonal H1N1, H3N2 lineages, the prevalent European swine H1N1, H1N2, and H3N2 sublineages and 18 reassortants. Genotyping was done according to Lu et al. 2007

parental human H3N2 virus was clearly distinct from that of the previous German strain (Fig. 2c, d; see also Schrader and Süss 2004). In most European countries the seroprevalence in 2002/2003 of the human-like swine H3N2 is lower than that of avian-like H1N1. Human-like swine H3N2 is (almost) absent in Poland and the Czech Republic, very low in Ireland (4.2 %), and below 60 % in Belgium and Germany. Only in Italy and Spain H3N2 prevalences are as high as H1N1 prevalences (Van Reeth et al. 2008; Simon-Grifé et al. 2010).

6.4 Emergence of Human-Like H1N2 in European Pigs

Swine H1N2 viruses that became prevalent in Europe were first isolated in Great Britain in 1994 (Brown et al. 1995; 1998). Available sequence data indicate that these H1N2 viruses resulted from repeated reassortment events involving a seasonal A/Chile/1/1983-like H1N1 virus (donor of HA) and a seasonal H3N2 virus (donor of NA) (Fig. 2a, d). Apparently, human H3N2 viruses circulated in pigs unrecognized for several years, as one member of this clade was already isolated in 1991 (A/swine/UK/119404/1991) (compare Zell et al. 2008b, therein Fig. 1b). Since the avian-like swine H1N1, but no human-like H3N2 viruses, circulated among British pig during that time, it has to be concluded that the former viruses were the donor of the internal segments. Three years later, the human-like swine H1N2 viruses spread to the European mainland: France (1997), Italy (1998), Belgium (1999), and Germany (2000) (Marozin et al. 2002; Van Reeth et al. 2000; Schrader and Süss 2003). In 2002/2003, the seroprevalences of H1N2 in Belgium and Spain exceeded that of human-like H3N2; it was low in Germany (32.1 %) and Italy (13.8 %) and very low in the Czech Republic (3 %) and Ireland (0.6 %) (Van Reeth et al. 2008).

The evolution of the three prevalent sublineages of the European swine influenza viruses is schematically depicted in Fig. 3.

6.5 Other Reassortant Swine Influenza Viruses Isolated in Europe

Sooner or later co-circulation of two or more influenza virus types within a population leads to reassortant viruses, but such reassortants may have little chance to replace either parent virus. The three prevalent European swine influenza viruses gave rise to three groups of reassortants. The first group comprises reassortants of seasonal human H3N2 and swine influenza viruses (Fig. 4). The strains A/swine/Potsdam/35/1982, A/swine/Karrenzien/2/1987, and A/swine/Leipzig/145/1992 (Schrader and Süss 2004) are examples of swine H3N2 viruses which emerged independently of each other in Germany. They have the six internal gene segments

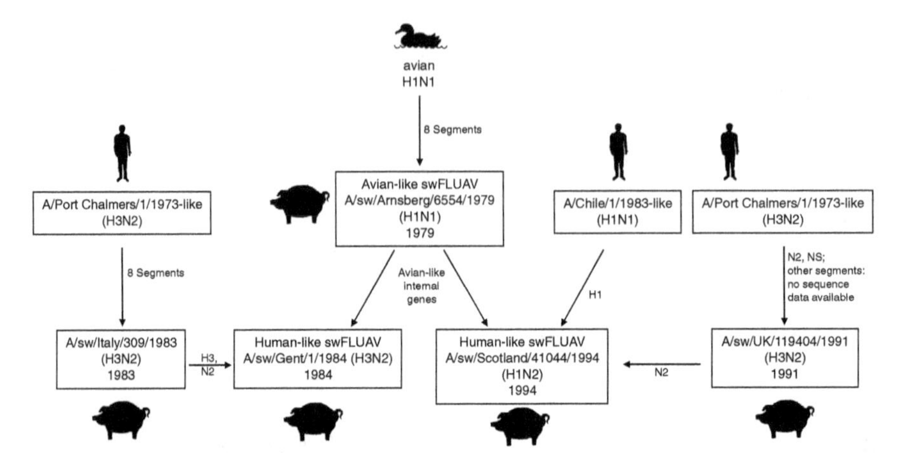

Fig. 3 Evolution of three prevalent sublineages of European swine influenza viruses

of avian-like H1N1 swine viruses and human HA and NA genes. These genes, however, branch independently of A/swine/Gent/1/1984-like viruses in phylogenetic trees (compare Fig. 2c, d) and are evidence of repeated 2 + 6 reassortments in pigs. Although only partial sequence data are available, further strains e.g., A/swine/Jena/3765/1989 (H3N2), A/swine/Leipzig/663/1992 (H3N2), and A/swine/Leipzig/318/1993 (H3N2), belong to this group and indicate that such reassortants may have circulated for 2–3 years. The second group of reassortants emerged in Italy. The preliminary characterization revealed a 7 + 1 reassortment between human H3N2 and swine H1N2 influenza viruses. These viruses have a neuraminidase gene of seasonal H3N2 viruses and seven segments (HA, internal genes) of human-like swine H1N2 viruses (Chiapponi et al. 2007). They circulated in Italy between 2003 and 2006.

The third group comprises reassortants between the prevalent sublineages of European swine influenza viruses. The three sublineages allow six HA/NA combinations and all of them have been detected in recent years. The compilation of Fig. 4 illustrates that several of these reassortments occurred repeatedly at different places and times: Some of the reassortants were published (Gourreau et al. 1994; Balint et al. 2009; Zell et al. 2008a, b), for others only preliminary reports are available (Chiapponi et al. 2007; Franck et al. 2007; Hjulsager et al. 2006). Such reassortants do not constitute antigenic shift mutants and failed to establish persistent stable infection chains yet.

Another rather unusual reassortant was isolated from pigs in England (Brown et al. 1994). The strain A/swine/England/191974/1992 (H1N7) was reported to comprise six segments of a human H1N1 virus (PB2, PB1, PA, HA, NP, NS) and the NA and M segments of an equine H7N7 virus (Brown et al. 1997a). Sequence data of the HA, NP, NA, and M segments are available in the GenBank. Although this virus represents an interesting reassortment, it has to be considered with some caution as the NA and M genes have a striking similarity to A/equine/Prague/1/

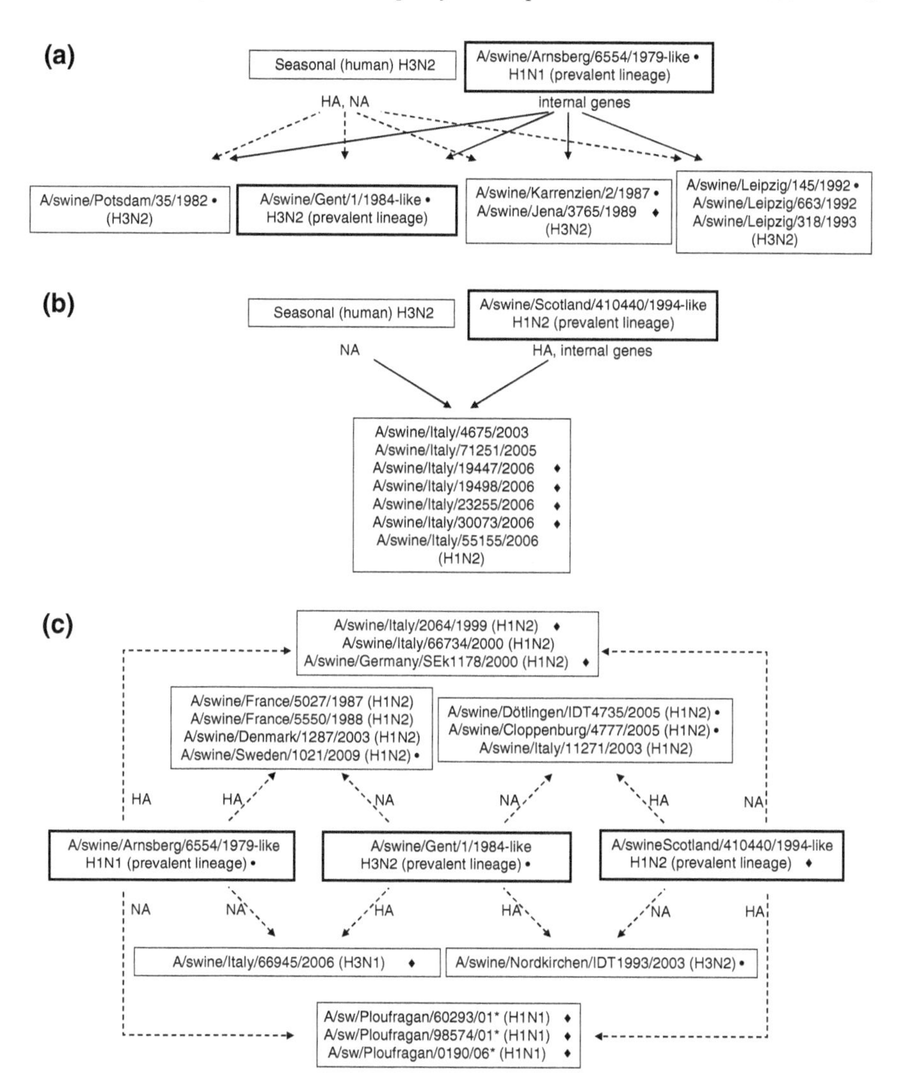

Fig. 4 Three groups of reassortant swine influenza viruses emerged in Europe. **a** 6 + 2 reassortants of avian-like swine H1N1 and human H3N2 influenza viruses. **b** 7 + 1 reassortants of human-like swine H1N2 and human H3N2 viruses. **c** Six different reassortants of the three prevalent European swine influenza virus lineages. The prevalent swine influenza virus lineages are boxed with bold lines. *Filled circles* (•) indicate the availability of complete sequence data in the GenBank, *filled diamonds* (♦) indicate partial sequence data. * Strains designations follow those of Franck et al. (2007); these strains were renamed when their sequences were deposited in the GenBank

1956 (H7N7). All other equine H7N7 sequences available from the GenBank (isolates of 1966–1977) show synonymous substitutions as a consequence of genetic drift and therefore differ significantly from A/equine/Prague/1/1956.

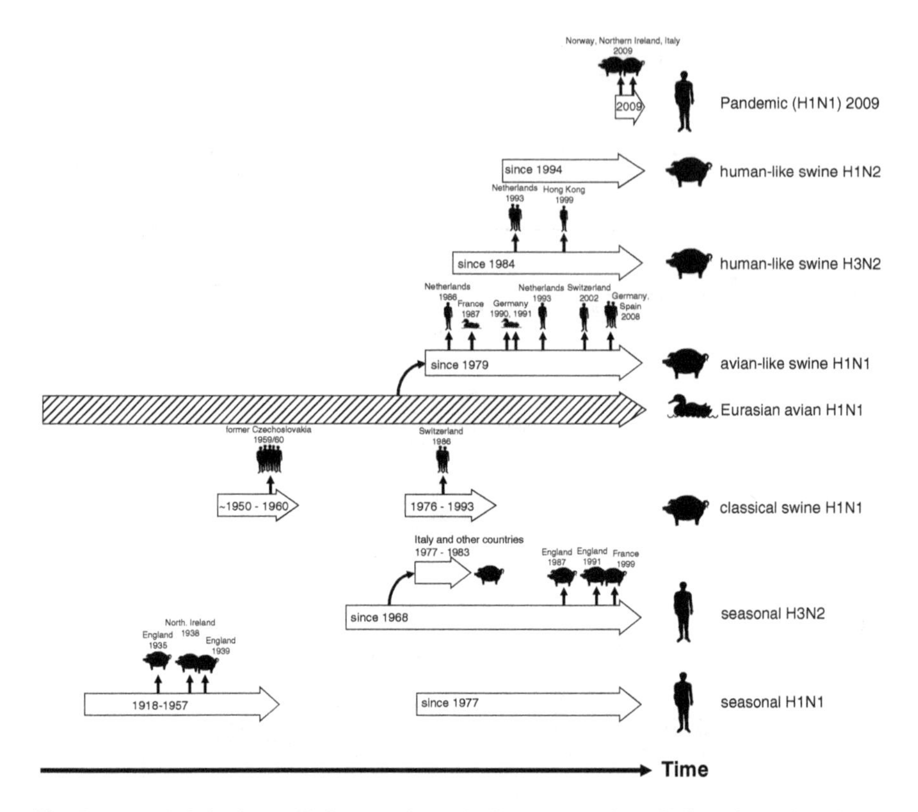

Fig. 5 Zoonotic infections of influenza viruses in Europe over time. Indicated are the relevant lineages of human, avian, swine influenza viruses in Europe and characteristic virus isolates

Equine H7N7 viruses disappeared around 1977 and it is quite astounding that an A/equine/Prague/1/1956-like virus should have persisted in an unknown reservoir for 36 years without accumulation of synonymous substitutions. Therefore, the biological significance of this reassortant should be scrutinized.

7 Zoonotic Infections

7.1 Human→Swine Infections

Human H1N1 influenza viruses have only a limited capacity to productively infect pigs (Hinshaw et al. 1978). However, there are several swine isolates of human origin from the 1930s which were isolated from clinically ill pigs (Lamont 1938; Blakemore and Gledhill 1941). Serological similarity of these strains to contemporary human strains was already observed by Glover (1941). Evidence for zoonotic infections of pigs by human H1N1 was also presented by Shope (1938). Thereafter, human–swine infections with H1N1 have not been documented in Europe, especially not after reemergence of H1N1 in 1977 (Fig. 5). However, there is indirect evidence that such infections may have occurred: (i) the emergence of H1N2 reassortants in swine in Europe (Brown et al. 1998; Marozin et al. 2002), and (ii) the observed antibody prevalence to human H1N1 in pig sera (Aymard et al. 1980). In Japan and China, several studies demonstrate the transmission of seasonal H1N1 to pigs as shown by virus isolation and seroprevalence studies (Goto et al. 1988; Katsuda et al. 1995; Nerome et al. 1982; Yu et al. 2007).

The human H2N2 viruses have never been isolated from pigs after natural infection, although there is one study that showed antibodies against H2N2 in four pigs in the former Czechoslovakia (Kaplan and Payne 1959). In principle, pigs are susceptible to these viruses as experimental infection of pigs with A/Singapore/1/1957 (H2N2) was successful (Patocka et al. 1958).

Seasonal H3N2 viruses were frequently detected in pigs in Europe and elsewhere. This is documented in several serological studies from Germany, the UK, France, Romania, and the Czech Republic (Sandow and Wildfuhr 1970; Harkness et al. 1972; Popovici et al. 1972; Aymard et al. 1980; Tumova et al. 1980; Pospisil et al. 2001). Occasionally, human H3N2 viruses were isolated from pigs (A/swine/England/163266/1987, A/swine/United Kingdom/119404/1991; Brown et al. 1998). In Italy, A/England/42/1972-like H3N2 viruses persisted from 1977–1983 in pigs (Ottis et al. 1982; Castrucci et al. 1993). Moreover, zoonotic infections with seasonal H3N2 gave rise to numerous reassortant viruses: the European human-like swine H3N2 lineage and the American tripple reassortants are the most prominent representatives (Castrucci et al. 1993; Olsen 2002; Karasin et al. 2000c). Figure 5 shows a compilation of zoonotic influenza virus infections in Europe.

The pandemic (H1N1) 2009 virus was repeatedly transmitted to pigs, first in Canada (Pasma and Joseph 2010), later in Norway, Northern Ireland, Italy (Hofshagen et al. 2009; Welsh et al. 2010; Moreno et al. 2010), and other European countries. The available reports indicate that this new reassortant induces a mild disease (Brookes et al. 2010; Itoh et al. 2009; Lange et al. 2009) and that many infections may be unrecognized. In addition, pandemic (H1N1) 2009 virus exhibits a significant cross-reaction to antibodies against avian-like swine H1N1 which impedes serological distinction (Kyriakis et al. 2010; Dürrwald et al. 2010). On the other hand, this cross-reactivity may hinder establishment of the pandemic virus in regions with high prevalence of avian-like swine H1N1.

7.2 Bird→Swine Infections

Although an initial bird→swine infection gave rise to the avian-like swine H1N1 sublineage, no discrete infections with avian influenza viruses have been documented in Europe. A similar observation was previously made in North America, when the genetic origin of 73 swine isolates (1976–1990) was investigated and no entry of avian genes could be detected (Wright et al. 1992). Later studies, however, revealed several of such infections in Asia, America, and Africa (Guan et al. 1996; El-Sayed et al. 2010; Karasin et al. 2000a, b; 2004; Yu et al. 2007, 2008; Peiris et al. 2001; Lee et al. 2009).

7.3 Swine→Bird Infections

In the 1980s and early 1990s, at least three swine-origin strains were isolated from birds (Andral et al. 1985; Ludwig et al. 1994; Wood et al. 1997). Partial genetic analyses revealed a reintroduction of avian-like swine H1N1 viruses into turkey farms (Ludwig et al. 1994). Improved hygiene in poultry husbandry and advanced adaptation of the swine H1N1 to its pig host may explain the failure of virus isolation in recent years in Europe. Human-like swine H3N2 and H1N2 strains have not been isolated from birds yet. Absence of α-2,6-linked sialic acids in poultry may be the main reason for the inability of human-like H3N2 and H1N2 sublineages to replicate in birds. The chapter by Yassine and colleagues on "Interspecies transmission of Influenza A viruses between swine and poultry" in this book described these interspecies infections in more detail.

7.4 Swine→Wild Boar Infections

In principle, wild boars should be susceptible to influenza viruses of swine and avian origin and may serve as a reservoir for such viruses. Although they have contacts to feral birds, the possibility of a transmission of avian influenza viruses to feral pigs is only insufficiently investigated in Europe. However, several serological studies searched for antibodies to swine influenza viruses in wild boars (recently reviewed in Kuntz-Simon and Madec 2009). Antibodies to avian-like swine H1N1 influenza viruses in feral pigs were detectable in Spain, Poland, and Croatia but not in Slovenia, Russia, and Ukraine. Another recent study demonstrated antibodies to avian-like swine H1N1 and human-like swine H3N2 viruses in Germany (Kaden et al. 2008). Two virus isolates described in that study [A/wild boar/WS169/2006 (H3N2), A/wild boar/WS188/2006 (H3N2)] should be considered with caution. The sequences of both isolates are identical and the published sequences of five different gene segments (HA, NP, NA, M, NS) show a sequence identity of nearly 100 % to A/swine/Bakum/909/1993 (H3N2) which was used as a H3N2 control in this study. Since influenza viruses exhibit a genetic drift due to the accumulation of synonymous and non-synonymous substitutions, one would expect some genetic variation in the course of several hundred virus generations (1993–2006). The wild boar isolates obviously lack this genetic drift.

Table 2 Zoonotic infection

No.	Country	Years	No. of patients	Type	Designation	Reference
1	Former Czechoslovakia	1959	6	Classical swine H1N1	No information available	Kluska et al. (1960)
2	Switzerland	1986	2	Avian-like swine H1N1	No information available	de Jong et al. (1986)
3	The Netherlands	1986	1	Avian-like swine H1N1	A/Netherlands/386/1986	de Jong et al. (1986); Rimmelzwaan et al. (2001)
4	The Netherlands	1993	1	Avian-like swine H1N1	A/Netherlands/477/1993	Rimmelzwaan et al. (2001)
5	The Netherlands	1993	2	Human-like swine H3N2	A/Netherlands/5/1993, A/Netherlands/35/1993	Claas et al. (1994)
6	Hong Kong	1999	1	Human-like swine H3N2	A/Hong Kong/1774/1999	Gregory et al. (2001)
7	Switzerland	2002	1	Avian-like swine H1N1	A/Switzerland/8808/2002	Gregory et al. (2003)
8	Germany	2007	1	Avian-like swine H1N1	A/Niedersachsen/58/2007	Schweiger et al. (2008)
9	Spain	2008	1	Avian-like swine H1N1	A/Aragon/RR3218/2008	Adiego Sancha et al. (2009)

7.5 Swine→Human Infections

Swine-to-human transmissions of classical swine H1N1 influenza viruses were first observed in Czechoslovakia in the 1950s (Kluska et al. 1961). Since then, sporadic infections were repeatedly demonstrated by virus isolation in the United States, Europe, and the Asian part of the former Soviet Union (reviewed in Myers et al. 2007). Several incidents of human infection with the European avian-like H1N1 and human-like H3N2 swine influenza viruses have been reported so far (Table 2) (Adiego Sancho et al. 2009; Claas et al. 1994; de Jong et al. 1986; Gregory et al. 2001, 2003; Rimmelzwaan et al. 2001; Schweiger et al. 2008). Apparently, zoonotic infections with the European swine viruses cause a benign disease with mild flu-like symptoms, whereas infections with classical swine strains may lead to more serious symptoms—few fatalities after infections with the latter viruses were reported (Myers et al. 2007). Despite repeated isolation of swine influenza viruses from human specimens, the prevalence of zoonotic infections in Europe is largely obscure. Previous work demonstrated seropositivity of personnel having contact to diseased pigs (Aymard et al. 1980; Sinnecker et al. 1983). A recent study conducted in Thuringia, Germany, indicates that approximately 15 % of the investigated sera of occupationally exposed humans (pig farmers, slaughterers, veterinarians) exhibit antibodies to the European lineages of swine influenza viruses (Krumbholz et al. 2010).

References

Adiego Sancho B, Omenaca Teres M, Martinez Cuenca S et al (2009) Human case of swine influenza A (H1N1), Aragon, Spain, November 2008. Eurosurveillance 14:1–2

Andral B, Toquin D, Madec F, Aymard M, Gourreau JM, Kaiser C, Fontaine M, Metz MH (1985) Disease in turkeys associated with H1N1 influenza virus following an outbreak of the disease in pigs. Vet Rec 116:617–618

Anhlan D, Grundmann N, Makalowski W, Ludwig S, Scholtissek C (2011) Origin of the 1918 pandemic H1N1 influenza A virus as studied by codon usage patterns and phylogenetic analysis. RNA 17:64–73

Aymard M, Brigaud M, Chastel C, Fontaine M, Tillon JP, Vannier P (1980) Comparison of influenza antibody serologic immunity in man and in pig. Comp Immunol Microbiol Infect Dis 3:111–119

Balint A, Metreveli G, Widen F, Zohari S, Berg M, Isaksson M, Renström LHM, Wallgren P, Belak S, Segall T, Kiss I (2009) The first Swedish H1N2 swine influenza virus isolate represents an uncommon reassortant. Virol J 6:180

Biront P, Meulemans G, Charlier G, Castrijck F (1980) Isolation of an influenza A virus related to the New Jersey strain (Hsw1N1) in fattening pigs. Vlaams Diergeneeskundig Tijdschrift 49:8–11

Blakemore F, Gledhill AW (1941) Discussion on swine influenza in the British Isles. Proc R Soc Med 34:611–615

Brookes SM, Nunez A, Choudhury B et al (2010) Replication, pathogenesis and transmission of pandemic (H1N1) 2009 virus in non-immune pigs, PLOS One 5:e9068

Brown IH (2000) The epidemiology and evolution of influenza viruses in pigs. Vet Microbiol 74:29–46

Brown IH, Alexander DJ, Chakraverty P, Harris PA, Manvell RJ (1994) Isolation of an influenza A virus of unusual subtype (H1N7) from pigs in England, and the subsequent experimental transmission from pig to pig. Vet Microbiol 39:125–134
Brown IH, Chakraverty P, Harris PA, Alexander DJ (1995) Disease outbreaks in pigs in Great Britain due to an influenza A virus of H1N2 subtype. Vet Rec 136(13):328–329
Brown IH, Hill ML, Harris PA, Alexander DJ, McCauley JW (1997a) Genetic characterisation of an influenza A virus of unusual subtype (H1N7) isolated from pigs in England. Arch Virol 142:1045–1050
Brown IH, Ludwig S, Olsen CW, Hannoun C, Scholtissek C, Hinshaw VS, Harris PA, McCauley JW, Strong I, Alexander DJ (1997b) Antigenic and genetic analyses of H1N1 influenza A viruses from European pigs. J Gen Virol 78:553–562
Brown IH, Harris PA, McCauley JW, Alexander DJ (1998) Multiple genetic reassortment of avian and human influenza A viruses in European pigs, resulting in the emergence of an H1N2 virus of novel genotype. J Gen Virol 79:2947–2955
Bush RM, Bender CA, Subbarao K, Cox NJ, Fitch WM (1999a) Predicting the evolution of human influenza A. Science 286:1921–1925
Bush RM, Fitch WM, Bender CA, Cox NJ (1999b) Positive selection on the H3 hemagglutinin of human influenza virus A. Mol Biol Evol 16(11):1457–1465
Campitelli L, Donatelli I, Foni E, Castrucci MR, Fabiani C, Kawaoka Y, Krauss S, Webster RG (1997) Continued evolution of H1N1 and H3N2 influenza viruses in pigs in Italy. Virology 232:310–318
Castro JM, del Pozo M, Simarro I (1988) Identification of H3N2 influenza virus isolated from pigs with respiratory problems in Spain. Vet Rec 122:418–419
Castrucci MR, Donatelli I, Sidoli L, Barigazzi G, Kawaoka Y, Webster RG (1993) Genetic reassortment between avian and human influenza A viruses in Italian pigs. Virology 193:503–506
Chiapponi C, Barbieri I, Manfredi R, Zanni I, Barigazzi G, Foni E (2007) Genetic diversity among H1N1 and H1N2 swine influenza viruses in Italy: preliminary results. In: Markowska-Daniel I (ed) 5th international symposium on emerging and re-emerging pig diseases. Krakow, Poland, p 261
Chu CM, Dawson IM, Elford WJ (1949) Filamentous forms associated with newly isolated influenza virus. Lancet 253:602–603
Claas ECJ, Kawaoka Y, De Jong JC, Masurel N, Webster RG (1994) Infection of children with avian-human reassortant influenza virus from pigs in Europe. Virology 204:453–457
Compans RW, Content J, Duesberg PH (1972) Structure of the ribonucleoprotein of influenza virus. J Virol 10(4):795–800
de Jong JC, de Ronde-Verloop JM, Bangma PJ, van Kregten E, Kerckhaert J, Paccaud MF, Wicki F, Wunderli W (1986) Isolation of swine-influenza-like A(H1N1) viruses from man in Europe. Lancet 328(8519):1329–1330
Duesberg PH (1968) The RNA's of influenza virus. Proc Natl Acad Sci USA 59:930–937
Dürrwald R, Krumbholz A, Baumgarte S, Schlegel M, Vahlenkamp TW, Selbitz HJ, Wutzler P, Zell R (2010) Swine influenza A vaccines, pandemic (H1N1) 2009 virus, and cross-reactivity. Emerg Inf Dis 16:1029–1030
Elkeles G (1934) Experimentelle Untersuchungen zur Aetiologie der Influenza (in German). Mededeelingen uit het Instituut voor Praeventieve Geneeskunde 1934:60–79
El-Sayed A, Awad W, Fayed A, Hamann HP, Zschöck M (2010) Avian influenza prevalence in pigs. Egypt. Emerg Inf Dis 16:726–727
Franck N, Queguiner S, Gorin S et al (2007). Molecular epidemiology of swine influenza virus in France: identification of novel H1N1 reassortants. In: Markowska-Daniel I (ed) 5th international symposium on emerging and re-emerging pig diseases, Krakow, Poland, p 250, 24–27 June 2007
Glover RE (1941) Discussion on swine influenza in the British Isles. Proc R Soc Med 34:615–617
Gorman OT, Bean WJ, Kawaoka Y, Donatelli I, Guo Y, Webster RG (1991) Evolution of influenza A virus nucleoprotein genes: implications for the origins of H1N1 human and classical swine viruses. J Virol 65(7):3704–3714

Goto H, Ogawa Y, Hirano T, Miwa Y, Piao FZ, Takai M, Noro S, Sakurada N (1988) Antibody responses of swine to type A influenza viruses during the past ten years in Japan. Epidemiol Infect 100:523–526
Gourreau JM, Kaiser C, Hannoun C, Vaissaire J, Gayot G (1980) First isolation in France of swine influenza virus (Hsw1N1) from a disease outbreak involving different microorganisms. Bulletin de l'Academie Veterinaire de France 53:181–188
Gourreau JM, Hannoun C, Kaiser C (1981) Diffusion du virus de la grippe du porc (Hsw1 N1) en France [in French]. Ann Virol (Inst Pasteur) 132E:287–294
Gourreau JM, Kaiser C, Valette M, Douglas AR, Labie J, Aymard M (1994) Isolation of two H1N2 influenza viruses from swine in France. Arch Virol 135:365–382
Gregory V, Lim W, Cameron K et al (2001) Infection of a child in Hong Kong by an influenza A H3N2 virus closely related to viruses circulation in European pigs. J Gen Virol 82:1397–1406
Gregory V, Bennett M, Thomas Y, Kaiser L, Wunderli W, Matter H, Hay A, Lin YP (2003) Human infection by a swine influenza A (H1N1) virus in Switzerland. Arch Virol 148:793–802
Guan Y, Shortridge KF, Krauss S, Li PH, Kawaoka Y, Webster RG (1996) Emergence of avian H1N1 influenza viruses in pigs in China. J Virol 70(11):8041–8046
Haesebrouck F, Pensaert M (1988) Influenza in swine in Belgium (1969–1986): epizootiologic aspects. Comp Immunol Microbiol Infect Dis 11:215–222
Haesebrouck F, Biront P, Pensaert MB, Leunen J (1985) Epizootics of respiratory tract disease in swine in Belgium due to H3N2 influenza virus and experimental reproduction of disease. Am J Vet Res 46:1926–1928
Harkness JW, Schild GC, Lamont PH, Brand CM (1972) Studies on relationships between human and porcine influenza. Bull WHO 46:709–719
Harnach R, Hubik R, Chivatal O (1950) Isolation of influenza virus in Czechoslovakia. Casopis Ceskoslavenskych Veterinaru 5:289
Hensley SE, Das SR, Bailey AL, Schmidt LM, Hickman HD, Jayaraman A, Viswanathan K, Raman R, Sasisekharan R, Bennink JR, Yewdell JW (2009) Hemagglutinin receptor binding avidity drives influenza A virus antigenic drift. Science 326:734–736
Hinshaw VS, Bean Jr WJ, Webster RG, Easterday BC (1978) The prevalence of influenza viruses in swine and the antigenic and genetic relatedness of influenza viruses from man and swine. Virology 84:51–62
Hjulsager CK, Bragstad K, Botner A, Nielsen EO, Vigre H, Enoe C, Larsen LE (2006). New swine influenza A H1N2 reassortment found in Danish swine. In: Proceedings of the 19th IPVS congress, Copenhagen. Abstract No. 0.55–03, p 265
Hofshagen M, Gjerset B, Er C et al (2009) Pandemic influenza A(H1N1)v: human to pig transmission in Norway? Eurosurveillance 14:pii:19406
Holland J, Spindler K, Horodyski F, Grabau E, Nichol S, VandePol S (1982) Rapid evolution of RNA genomes. Science 215:1577–1585
Ito T, Kawaoka Y (2000) Host-range barrier of influenza A viruses. Vet Microbiol 74:71–75
Ito T, Couceiro JNSS, Kelm S, Baum LG, Krauss S, Castrucci MR, Donatelli I, Kida H, Paulson JC, Webster RG, Kawaoka Y (1998) Molecular Basis for the generation in pigs of influenza A viruses with pandemic potential. J Virol 72(9):7367–7373
Itoh Y, Shinya K, Kiso M, Watanabe T, Sakoda Y, Hatta M, Muramoto Y, Tamura D, Sakai-Tagawa Y, Noda T, Sakabe S, Imai M, Hatta Y, Watanabe S, Li C, Yamada S, Fujii K, Murakami S, Imai H, Kakugawa S, Ito M, Takano R, Iwatsuki-Horimoto K, Shimojima M, Horimoto T, Goto H, Takahashi K, Makino A, Ishigaki H, Nakayama M, Okamatsu M, Takahashi K, Warshauer D, Shult PA, Saito R, Suzuki H, Furuta Y, Yamashita M, Mitamura K, Nakano K, Nakamura M, Brockman-Schneider R, Mitamura H, Yamazaki M, Sugaya N, Suresh M, Ozawa M, Neumann G, Gern J, Kida H, Ogasawara K, Kawaoka Y (2009) In vitro and in vivo characterization of new swine-origin H1N1 influenza viruses. Nature 460:1021–1025
Kaden V, Lange E, Starick E, Brue W, Krakowski W, Klopries M (2008) Epidemiological survey of swine infuenza A virus in selected wild boar populations in Germany. Vet Microbiol 131:123–132

Kaplan MM, Payne AMM (1959) Serological survey in animals for type A influenza in relation to the 1957 pandemic. Bull WHO 20:465–488
Karasin AI, Brown IH, Carman S, Olsen CW (2000a) Isolation and characterization of H4N6 avian influenza viruses from pigs with pneumonia in Canada. J Virol 74:9322–9327
Karasin AI, Olsen CW, Brown IH, Carman S, Stalker M, Josephson G (2000b) H4N6 influenza virus isolated from pigs in Ontario. Can Vet J 41:938–939
Karasin AI, Schutten MM, Cooper LA, Smith CB, Subbarao K, Anderson GA, Carman S, Olsen CW (2000c) Genetic characterization of H3N2 influenza viruses isolated from pigs in North America, 1977–1999: evidence for wholly human and reassortant virus genotypes. Virus Res 68:71–85
Karasin AI, West K, Carman S, Olsen CW (2004) Characterization of avian H3N3 and H1N1 influenza A viruses isolated from pigs in Canada. J Clin Microbiol 42:4349–4354
Karlas A, Nachuy N, Shin Y, Pleissner KP, Artarini A, Heuer D, Becker D, Khalil H, Ogilvie LA, Hess S, Mäurer AP, Müller E, Wolff T, Rudel T, Meyer TF (2010) Genome-wide RNAi screen identifies human host factors crucial for influenza virus replication. Nature 463:818–822
Katsuda K, Sato S, Shirahata T, Lindstrom S, Nerome R, Ishida M, Nerome K, Goto H (1995) Antigenic and genetic characteristics of H1N1 human influenza virus isolated from pigs in Japan. J Gen Virol 76:1247–1249
Kawaoka Y, Cox NJ, Haller O et al (2005) Family orthomyxoviridae. In: Fauquet CM (eds). Virus taxonomy. Eight report of the international committee on taxonomy of viruses, Elsevier Academic Press, Amsterdam, pp 681–693
Kilbourne ED (1968) Recombination of influenza A viruses of human and animal origin. Science 160:74–76
Kluska V, Macku M, Mensik J (1961) Demonstration of antibodies against swine influenza viruses in man (in Czech). Cesk Pediatr 16:408–414
Köbe K (1933) Die Aetiologie der Ferkelgrippe (enzootische Pneumonie des Ferkels) (in German). Zentralbl Bakt 1. Abt 129:161–176
Koen JS (1919) A practical method for field diagnosis of swine diseases. Am J Vet Med 14: 468–470
König R, Stertz S, Zhou Y, Inoue A, Hoffmann HH, Bhattacharyya S, Alamares JG, Tscherne DM, Ortigoza MB, Liang Y, Gao Q, Andrews SE, Bandyopadhyay S, De Jesus P, Tu BP, Pache L, Shih C, Orth A, Bonamy G, Miraglia L, Ideker T, García-Sastre A, Young JA, Palese P, Shaw ML, Chanda SK (2010) Human host factors required for influenza virus replication. Nature 463:813–817
Krauss S, Obert CA, Franks J, Walker D, Jones K, Seiler P, Niles L, Pryor SP, Obenauer JC, Naeve CW, Widjaja L, Webby RJ, Webster RG (2007) Influenza in migratory birds and evidence of limited intercontinental virus exchange. PLoS Pathog 3:e167
Krumbholz A, Schmidtke M, Bergmann S, Motzke S, Bauer K, Stech J, Dürrwald R, Wutzler P, Zell R (2009) High prevalence of amantadine resistance among circulating European porcine influenza A viruses. J Gen Virol 90:900–908
Krumbholz A, Lange J, Dürrwald R, Hoyer H, Bengsch S, Wutzler P, Zell R (2010) Prevalence of antibodies to swine influenza viruses in humans with occupational exposure to pigs, Thuringia, Germany, 2008–2009. J Med Virol 82:1617–1625
Kuiken T, Holmes EC, McCauley J, Rimmelzwaan GF, Williams CS, Grenfell BT (2006) Host species barriers to influenza virus infections. Science 312:394–397
Kuntz-Simon G, Madec F (2009) Genetic and antigenetic evolution of swine influenza viruses in Europe and evaluation of their zoonotic potential. Zoonoses Public Health 56:310–325
Kyriakis CS, Olsen CW, Carman S, Brown IH, Brookes SM, Van Doorsselaere J, Van Reeth K (2010) Serologic cross-reactivity with pandemich (H1N1) 2009 virus in pigs, Europe. Emerg Inf Dis 16:96–99
Lamont HG (1938) The problems of the practitioner in connection with the differential diagnosis and treatment of diseases of young pigs. Vet Rec 50:1377
Lange E, Kalthoff D, Blohm U, Teifke JP, Breithaupt A, Maresch C, Starick E, Fereidouni S, Hoffmann B, Mettenleiter TC, Beer M, Vahlenkamp TW (2009) Pathogenesis and transmission of the novel swine-origin influenza virus A/H1N1 after experimental infection of pigs. J Gen Virol 90:2119–2123

Lee JH, Pascua PN, Song MS, Baek YH, Kim CJ, Choi HW, Sung MH, Webby RJ, Webster RG, Poo H, Choi YK (2009) Isolation and genetic characterization of H5N2 influenza viruses from pigs in Korea. J Virol 83(9):4205–4215

Loeffen WL, Kamp EM, Stockhofe-Zurwieden N, van Nieuwstadt AP, Bongers JH, Hunneman WA, Elbers AR, Baars J, Nell T, van Zijderveld FG (1999) Survey of infectious agents involved in acute respiratory disease in finishing pigs. Vet Rec 145:123–129

Lu G, Rowley T, Garten R, Donis RO (2007) FluGenome: a web tool for genotyping influenza A virus. Nucl Acids Res 35:W275–W279

Ludwig S, Haustein A, Kaleta EF, Scholtissek C (1994) Recent influenza A (H1N1) infections of pigs and turkeys in Northern Europe. Virology 202:281–286

Madec F, Gourreau JM, Kaiser C, Aymard M (1984) Apparition de manifestations grippales chez les porcs en association avec un virus A/H3N2. Bull Acad Vet Fr 57:513–522

Makarova NV, Kaverin NV, Krauss S, Senne D, Webster RG (1999) Transmission of Eurasian avian H2 influenza virus to shorebirds in North America. J Gen Virol 80:3167–3171

Marozin S, Gregory V, Cameron K, Bennett M, Valette M, Aymard M, Foni E, Barigazzi G, Lin Y, Hay A (2002) Antigenic and genetic diversity among swine influenza A H1N1 and H1N2 viruses in Europe. J Gen Virol 83:735–745

Martinsson K, Klingeborn B, Rockborn G (1983) Utbrott av influenza suis i Sverige. (in Swedish). Svensk Vet Tidn 35:537

Moreno A, Di Trani L, Alborali L, Vaccari G, Barbieri I, Falcone E, Sozzi E, Puzelli S, Ferri G, Cordioli P (2010) First pandemic H1N1 outbreak from a pig farm in Italy. Open Virol J 4:52–56

Munster VJ, Baas C, Lexmond P, Waldenström J, Wallensten A, Fransson T, Rimmelzwaan GF, Beyer WEP, Schutten M, Olsen B, Osterhaus ADME (2007) Spatial, temporal, and species variation in prevalence of influenza A viruses in wild migratory birds. PLoS Path 3(5):e61

Myers KP, Olsen CW, Gray GC (2007) Cases of swine influenza in humans: a review of the literature. Clin Infect Dis 44:1084–1088

Nardelli L, Pascucci S, Gualandi GL, Loda P (1978) Outbreaks of classical swine influenza in Italy in 1976. Zentralbl Veterinarmed B 25:853–857

Nerome K, Ishida M, Oya A, Kanai C, Suwicha K (1982) Isolation of an influenza H1N1 virus from a pig. Virology 117:485–489

Neumeier E, Meier-Ewert H (1992) Nucleotide sequence analysis of the HA1 coding portion of the haemagglutinin gene of swine H1N1 influenza viruses. Virus Res 23:107–117

Neumeier E, Meier-Ewert H, Cox NJ (1994) Genetic relatedness between influenza A (H1N1) viruses isolated from humans and pigs. J Gen Virol 75:2103–2107

Nobusawa E, Sato K (2006) Comparison of the mutation rates of human influenza A and B viruses. J Virol 80(7):3675–3678

Noda T, Sagara H, Yen A, Takada A, Kida H, Cheng RH, Kawaoka Y (2006) Architecture of ribonucleoprotein complexes in influenza A virus particles. Nature 439:490–492

Olsen CW (2002) The emergence of novel swine influenza viruses in North America. Virus Res 85:199–210

Olsen B, Munster VJ, Wallensten A, Waldenstrom J, Osterhaus AD, Fouchier RA (2006) Global patterns of influenza A virus in wild birds. Science 312:384–388

Ottis K, Bachmann PA (1980) Occurrence of Hsw1N1 subtype influenza A viruses in wild ducks in Europe. Arch Virol 63:185–190

Ottis K, Bollwahn W, Bachmann PA, Heinritzi K (1981) Ausbruch von Schweineinfluenza in der Bundesrepublik Deutschland: Klinik, Nachweis und Differenzierung (in German). Tierärztl Umsch 36:608–612

Ottis K, Sidoli L, Bachmann PA, Webster RG, Kaplan MM (1982) Human influenza a viruses in pigs: isolation of a H3N2 strain antigenically related to A/England/42/72 and evidence for continuous circulation of human viruses in the pig population. Arch Virol 73:103–108

Parvin JD, Moscona A, Pan WT, Leider JM, Palese P (1986) Measurement of the mutation rates of animal viruses: influenza A virus and poliovirus type 1. J Virol 59(2):377–383

Pasma T, Joseph T (2010) Pandemic (H1N1) 2009 infection in swine herds, Manitoba, Canada. Emerg Inf Dis 16:706–708

Patocka F, Schreiber E, Kubelka V, Korb J, John C, Schön E (1958) An attempt ot transmit the human influenza virus strain A-Sing 57 to swine; preliminary report. J Hyg Epidemiol Microbiol Immunol 2(1):9–15
Peiris JS, Guan Y, Markwell D, Ghose P, Webster RG, Shortridge KF (2001) Cocirculation of avian H9N2 and contemporary human H3N2 influenza A viruses in pigs in southeastern China: potential for genetic reassortment? J Virol 75:9679–9686
Pensaert M, Ottis K, Vandeputte J, Kaplan MM, Bachmann PA (1981) Evidence for the natural transmission of influenza A virus from wild ducks to swine and its potential importance for man. Bull WHO 59(1):75–78
Popovici V, Hiastru F, Draghici D, Zilisteanu E, Matepiuc M, Cretescu L, Niculescu I (1972) Infection of pigs with an influenza virus related to the A2-Hong Kong-1-68 strain. Acta Virol 16:363
Pospisil Z, Lany P, Tumova B, Buchta J, Zendulkova D, Cihal P (2001) Swine influenza surveillance and the impact of human influenza epidemics on pig herds in the Czech Republic. Acta Vet Brno 70:327–332
Rimmelzwaan GF, de Jong JC, Bestebroer TM, van Loon AM, Claas ECJ, Fouchier RAM, Osterhaus ADME (2001) Antigenic and genetic characterization of swine influenza A (H1N1) viruses isolated from pneumonia patients in the Netherlands. Virology 282:301–308
Roberts DH, Cartwright SF, Wibberley G (1987) Outbreaks of classical swine influenza in pigs in England in 1986. Vet Rec 121:53–55
Rogers GN, Paulsen JC (1983) Receptor determinants of human and animal influenza virus isolates: differences in receptor specificity of the H3 hemagglutinin based on species of origin. Virology 127:361–373
Sandow D, Wildfuhr W (1970) Comparative serological studies of the influenza strains A2 Hongkong-1-68 and A2 DDR-Berlin in pigs. Monatsh Veterinärmed 25:320–322
Schmidtke M, Zell R, Bauer K, Krumbholz A, Schrader C, Süss J, Wutzler P (2006) Amantadine resistance among porcine H1N1, H1N2, and H3N2 influenza A viruses isolated in Germany between 1981 and 2001. Intervirology 49:286–293
Scholtissek C (1990) Pigs as "mixing vessels" for the creation of new pandemic influenza A viruses. Med Princ Pract 2:65–71
Scholtissek C, Bürger H, Bachmann PA, Hannoun C (1983) Genetic relatedness of hemagglutinins of the H1N1 subtype of influenza A viruses isolated from swine and birds. Virology 129:521–523
Scholtissek C, Bürger H, Kistner O, Shortridge KF (1985) The nucleoprotein as a possible major factor in determining host specificity of influenza H3N2 viruses. Virology 147:287–294
Schrader C, Süss J (2003) Genetic characterization of a porcine H1N2 influenza virus strain isolated in Germany. Intervirology 46:66–70
Schrader C, Süss J (2004) Molecular epidemiology of porcine H3N2 influenza A viruses isolated in Germany between 1982 and 2001. Intervirology 47:72–77
Schultz U, Fitch WM, Ludwig S, Mandler J, Scholtissek C (1991) Evolution of pig influenza viruses. Virology 183:61–73
Schweiger B, Heckler R, Biere B (2008) Characterization of a porcine influenza virus isolated from a human sample. In: Programme and abstracts, abstract CLV2, 18th Annual Meeting Gesellschaft für Virologie, 5–8 March 2008, Heidelberg
Shope RE (1931a) The etiology of swine influenza. Science 73:214–215
Shope RE (1931b) Swine influenza: I. Experimental transmission and pathology. J Exp Med 54:349–362
Shope RE (1938) Serological evidence for the occurrence of infection with human influenza virus in swine. J Exp Med 67:739–748
Shope RE, Francis T (1936) The susceptibility of swine to the virus of human influenza. J Exp Med 64:791–801
Simon-Grifé M, Martin-Valls GE, Vilar MJ (2010) Seroprevalence and risk factors of swine influenza in Spain. Vet Microbiol, (in press) (doi: 10.1016/j.vetmic.2010.10.015)

Sinnecker H, Sinnecker R, Zilske E, Strey A, Leopoldt D (1983) Influenzavirus A/swine-Ausbrüche bei Hausschweinen und Antikörperbefunde in Humanseren [in German]. Zentralbl Bakteriol Mikrobiol Hyg A 255(2–3):209–213
Stech J, Xiong X, Scholtissek C, Webster RG (1999) Independence of evolutionary and mutational rates after transmission of avian influenza viruses to swine. J Virol 73(3):1878–1884
Steinhauer DA (1999) Role of hemagglutinin cleavage for the pathogenicity of influenza virus. Virology 258:1–20
Subbarao EK, London W, Myrphy BR (1993) A single amino acid in the PB2 gene of influenza A virus is a determinant of host range. J Virol 67(4):1761–1764
Tamura K, Dudley J, Nei M, Kumar S (2007) MEGA4: molecular evolutionary genetics analysis (MEGA) software version 4.0. Mol Biol Evol 24:1596–1599
Tumova B, Stumpa A, Mensik J (1980) Surveillance of influenza in pig herds in Czechoslovakia in 1974–1979. 2. Antibodies against influenza A (H3N2) and A (H1N1) viruses. Zentralbl Veterinärmed B 27:601–607
Van Reeth K, Brown IH, Pensaert M (2000) Isolations of H1N2 influenza A virus from pigs in Belgium. Vet Rec 146(20):588–589
Van Reeth K, Brown IH, Dürrwald R, Foni E, Labarque G, Lenihan P, Maldonado J, Markowska-Daniel I, Pensaert M, Pospisil Z, Koch G (2008) Seroprevalence of H1N1, H3N2 and H1N2 influenza viruses in pigs in seven European countries in 2002–2003. Influenza Other Respi Viruses 2:99–105
Vandeputte J, Pensaert M, Castryck F (1980) Serological diagnosis and distribution of swine influenzavirus in Belgium. Vlaams Diergeneeskundig Tijdschrift 49:1–7
Wahlgren A, Waldenström J, Sahlin S, Haemig PD, Fouchier RAM, Osterhaus ADME, Pinhassi J, Bonnedahl J, Pisareva M, Grudinin M, Kiselev O, Hernandez J, Falk KI, Lundkvist A, Olsen B (2008) Gene segment reassortment between American and Asian lineages of avian influenza virus from waterfowl in the Beringia area. Vector-borne Zoon Dis 8(6):783–789
Waldmann O (1933) Die Aetiologie des Ferkelkümmerns. Die Ferkelgrippe. (in German). Berl Tierarztl Wochenschr 44:693–697
Wallensten A, Munster VJ, Elmberg J, Osterhaus ADME, Fouchier RAM, Olsen B (2005) Multiple gene segment reassortment between Eurasian and American lineages of influenza A virus (H6N2) in Guillemot (*Uria aalge*). Arch Virol 150:1685–1692
Webster RG, Bean WJ, Gorman WT, Chambers TM, Kawaoka Y (1992) Evolution and ecology of influenza A viruses. Microbiol Rev 56(1):152–179
Welsh MD, Baird PM, Guelbenzu-Gonzalo MP, Hanna A, Reid SM, Essen S, Russell C, Thomas S, Barrass L, McNeilly F, McKillen J, Todd D, Harkin V, McDowell S, Choudhury B, Irvine RM, Borobia J, Grant J, Brown IH (2010) Initial incursion of pandemic (H1N1) 2009 influenza A virus into European pigs. Vet Rec 166:642–645
Witte KK, Nienhoff H, Ernst H, Schmidt U, Prager D (1981) Erstmaliges Auftreten einer durch das Schweineinfluenzavirus verursachten Epizootie in Schweinebeständen der Bundesrepublik Deutschland [in German]. Tierärztliche Umschau 36(9):591–606
Wood GW, Banks J, Brown IH, Strong I, Alexander DJ (1997) The nucleotide sequence of the HA1 of the haemagglutinin of an HI avian influenza virus isolate from turkeys in Germany provides additional evidence suggesting recent transmission from pigs. Avian Pathol 26:347–355
Wright SM, Kawaoka Y, Sharp GB, Senne DA, Webster RG (1992) Interspecies transmission and reassortment of influenza A viruses in pigs and turkeys in the United States. Am J Epidemiol 136:488–497
Yewdell JW, Webster RG, Gerhard WU (1979) Antigenic variation in three distinct determinants of an influenza type A haemagglutinin molecule. Nature 279:246–248 (17 May 1979)
Yoshioka Y, Sugita S, Kanegae Y, Shortridge KF, Nerome K (1994) Origin and evolutionary pathways of the H1 haemagglutinin gene of avian, swine and human influenza viruses: cocirculation of two distinct lineages of swine virus. Arch Virol 134:17–28
Yu H, Zhang GH, Hua RH, Zhang Q, Liu TQ, Liao M, Tong GZ (2007) Isolation and genetic analysis of human origin H1N1 and H3N2 influenza viruses from pigs in China. Biochem Biophys Res Commun 356:91–96

Yu H, Hua RH, Wei TC, Zhou YJ, Tian ZJ, Li GX, Liu TQ, Tong GZ (2008) Isolation and genetic characterization of avian origin H9N2 influenza viruses from pigs in China. Vet Microbiol 131:82–92

Yus E, Sanjuan ML, Garcia F, Castro JM, Simarro I (1992) Influenza A viruses: epidemiologic study in fatteners in Spain (1987–1989). Zentralbl Veterinärmed B 39:113–118

Zell R, Krumbholz A, Eitner A, Krieg R, Halbhuber KJ, Wutzler P (2007) Prevalence of PB1-F2 of influenza A viruses. J Gen Virol 88:536–546

Zell R, Motzke S, Krumbholz A, Wutzler P, Herwig V, Dürrwald R (2008a) Novel reassortant of swine influenza H1N2 virus in Germany. J Gen Virol 89:271–276

Zell R, Bergmann S, Krumbholz A, Wutzler P, Dürrwald R (2008b) Ongoing evolution of swine influenza viruses: a novel reassortant. Arch Virol 153:2085–2092

Zhang XM, Herbst W, Lange-Herbst H, Schliesser T (1989) Seroprevalence of porcine and human influenza A virus antibodies in pigs between 1986 and 1988 in Hassia. Zentralbl Veterinärmed B 36:765–770

zu Dohna H, Li J, Cardona CJ, Miller J, Carpenter TE (2009). Invasions by Eurasian avian influenza virus H6 genes and replacement of the virus' North American clade. Emerg Inf Dis 15:1040–1045

History of Swine Influenza Viruses in Asia

Huachen Zhu, Richard Webby, Tommy T. Y. Lam, David K. Smith, Joseph S. M. Peiris and Yi Guan

Abstract The pig is one of the main hosts of influenza A viruses and plays important roles in shaping the current influenza ecology. The occurrence of the 2009 H1N1 pandemic influenza virus demonstrated that pigs could independently facilitate the genesis of a pandemic influenza strain. Genetic analyses revealed that this virus was derived by reassortment between at least two parent swine influenza viruses (SIV), from the northern American triple reassortant H1N2 (TR) and European avian-like H1N1 (EA) lineages. The movement of live pigs between different continents and subsequent virus establishment are preconditions for such a reassortment event to occur. Asia, especially China, has the largest human and pig populations in the world, and seems to be the only region frequently importing pigs from other continents. Virological surveillance revealed that not only classical swine H1N1 (CS), and human-origin H3N2 viruses circulated, but all of the EA, TR and their reassortant variants were introduced into and co-circulated in pigs in this region. Understanding the long-term evolution and history of SIV in Asia would provide insights into the emergence of influenza viruses with epidemic potential in swine and humans.

H. Zhu · T. T. Y. Lam · D. K. Smith · Y. Guan
International Institute of Infection and Immunity,
Shantou University Medical College, Shantou, Guangdong, China

H. Zhu · T. T. Y. Lam · D. K. Smith · J. S. M. Peiris · Y. Guan (✉)
State Key Laboratory of Emerging Infectious Diseases,
5/F, Li Ka Shing Faculty of Medicine, The University of Hong Kong,
21 Sassoon Road, Pokfulam, Hong Kong SAR, China
e-mail: yguan@hku.hk

R. Webby
Division of Virology, Department of Infectious Disease,
St. Jude Children's Research Hospital, Memphis, TN 38105, USA

Current Topics in Microbiology and Immunology (2013) 370: 57–68
DOI: 10.1007/82_2011_179

Published Online: 23 September 2011

Contents

1 Introduction

Asia is home to the world's largest human population. With its rapid growth and increasing wealth, Asia has an escalating need and demand for more and better quality food sources. Farming practices in Asia have been greatly changed by industrialization and globalization in the quest for greater production. Importation of breeding pigs from the USA and Europe and the establishment of intensive pig-breeding farms have caused the population of pigs in Asia to soar. Now, approximately 60% of the world's pigs are in Asia and China alone is home to over 40% of the world's pigs (USDA/FAS).

Greatly increased swine numbers have led to enhanced opportunities for contact between pigs and humans and between pigs and the similarly enlarged poultry flocks and the wild birds of Asia. Interspecies transmission of influenza to and from swine has been frequently observed (Pensaert et al. 1981; Mohan et al. 1981; Claas et al. 1994; Guan et al. 1996) and pigs are regarded as a major intermediate host in the process of adapting avian viruses to mammalian hosts (Scholtissek et al. 1985; Scholtissek 1990). Highly pathogenic H5N1 and low pathogenic H9N2 avian influenza viruses have become enzootic in Asia and transmissions to humans and pigs have occurred (Webster et al. 2006; Peiris et al. 1999; Lin et al. 2000; Claas et al. 1998; Subbarao et al. 1998). The 2009 pandemic virus had its origins in currently circulating swine influenza viruses (Dawood et al. 2009; Garten et al. 2009; Smith et al. 2009b).

Both the economic consequences to food production and the threat to human health emphasize the importance of swine influenza. Monitoring the evolution and ecology of this virus is an essential task for human well-being. This is especially the case in Asia where the largest population of pigs in the world interacts with such a large human population.

2 Prevalence and Detection of Swine Influenza Viruses in Asia

H1N1 and H3N2 influenza viruses occur in pigs in Asia but the clinical picture of infection is less clear. Pig farming in China and much of Asia has been traditionally based on small family holdings and only in the last 20 years or so have large-scale farming operations come onstream. Consequently, relatively little surveillance has occurred and there is limited information about swine influenza in Asia in the wider scientific literature.

2.1 Classical Swine H1N1 Virus

The presence of classical swine H1N1 (CS) influenza in China probably dates back to the 1918–1919 pandemic when, in the wake of human infections, an outbreak of high mortality occurred in pigs in cities along the Chinese coast (Chun 1919; Kilbourne 2006). A similar disease pattern was observed in the USA with humans infected before pigs (Koen 1919; Reid and Taubenberger 2003). Findings from evolutionary studies revealed that classical swine H1N1 and 1918 pandemic H1N1 viruses shared a common ancestor or were highly closely related to each other (Smith et al. 2009a; Kanegae et al. 1994; Gorman et al. 1991). This historic interrelationship between human and porcine H1N1 viruses may be similar to the current situation with H1N1/2009 viruses. Since its emergence, the 2009 human pandemic H1N1 virus has been repeatedly transmitted from humans to pigs (Pereda et al. 2010; Weingartl et al. 2010; Vijaykrishna et al. 2010).

Although the CS virus was isolated and identified as early as 1930 in the USA (Shope 1931), it was first isolated in Asia in 1974 (swine/Hong Kong/1/74). Since the mid-1970s influenza surveillance of pigs conducted in Hong Kong and Japan has revealed that classical H1N1 viruses are widely distributed in many Asian regions and countries. Surveillance in Hong Kong from 1976 to 1980 took samples from pigs grown in Hong Kong and pigs imported from Mainland China, Taiwan and Singapore (Shortridge and Webster 1979). Regular isolation of classical swine viruses showed their continuing presence in apparently healthy pigs (Shortridge and Webster 1979; Yip 1976). During the same period of time, serological studies revealed that classical swine H1N1 viruses were also common in the pig populations of Japan (Arikawa et al. 1979; Nerome et al. 1982; Ogawa et al. 1983; Yamane et al. 1978).

Further surveillance in Hong Kong during 1993–1994 showed that classical H1N1 viruses were apparently the predominant influenza virus infecting pigs (Guan Y, unpublished). A clear epizootic occurred, with large numbers of viruses isolated in February, March and April of 1994. Little surveillance has been conducted in other countries of Asia. Classical H1N1 swine influenza viruses were reported to be isolated in Thailand in 1988 (Kupradinun et al. 1991) and Mainland China in 1991 (Guan et al. 1996), and this virus was also identified from pigs in

Korea (Lee et al. 2008; Song et al. 2007) and India (Das et al. 1981; Chatterjee et al. 1995).

Generally, classical H1N1 viruses were genetically stable and showed minor antigenic drift in Asian countries. However, a reassortant H1N2 virus, with its N2 segment from early human-like H3N2 viruses and its remaining segments from classical H1N1 viruses (Sugimura et al. 1980), caused a major outbreak in southern Japan from winter 1989 to spring 1990. Affected pigs had a typical influenza illness and most swine tested possessed corresponding antibodies (Ouchi et al. 1996). Similar reassortant viruses were also detected in Hong Kong from pigs imported from China during 1999 to 2004 (Vijaykrishna et al. 2011).

Phylogenetic analyses of large data sets of swine influenza viruses reveal that the classical H1N1 viruses isolated in Hong Kong do not form a single monophyletic group. They are interspersed with North American CS viruses, indicating multiple introductions of CS into Asian countries from the USA (Vijaykrishna et al. 2011). However, it is hard to believe that the classical H1N1 viruses from different Asian countries were all introduced from the USA via importation of pigs from there. CS viruses might have evolved from the 1918 pandemic H1N1 virus as it became able to persist in pigs. Supposedly, then, this virus should have existed in all regions where pigs were available. However, the effect of long-term vaccination practices in the USA might induce an evolutionary advantage in CS viruses from that region, allowing them to replace previously existing CS strains. This hypothesis could be possibly tested again as the 2009 pandemic H1N1 virus and its variants become established in pigs.

2.2 H3N2 Human-like Influenza Viruses

H3N2 swine viruses appear to be the result of multiple transmissions of viruses from humans to pigs. They were first isolated in Asia from pigs in Taiwan soon after the Hong Kong pandemic (Kundin 1970). During surveillance in Hong Kong from 1976 to 1982, and from 1998 to the present, contemporary variants of H3N2 human-like viruses and antibodies to them were regularly isolated or detected in pigs from Asian countries (Shortridge and Webster 1979; Vijaykrishna et al. 2011; Shortridge et al. 1977; Webster et al. 1977). Based on our long-term surveillance, almost all of the major human H3N2 variants could be introduced into the pig population (Vijaykrishna et al. 2011). Some of these variants might remain in somewhat genetically dormant states, akin to evolutionary stasis, in pigs even many years after their counterparts had disappeared from humans (Shortridge et al. 1977).

The first two major H3N2 variants were A/Port Chalmers/1/73 (PC) and A/Victoria/3/75 (Vic) (Shortridge and Webster 1979; Shortridge et al. 1977, 1979). Interspecies transmission to pigs of these two viruses occurred in many Asian countries, including China, Korea and Japan (Shortridge and Webster 1979; Shortridge et al. 1977, 1979; Song et al. 2003; Jung and Song 2007; Yamane et al.

1979; Nerome et al. 1981; Arikawa et al. 1982). In Europe, PC-like H3N2 viruses reassorted with European avian (EA)-like H1N1 viruses to generate an H3N2 virus with PC-like surface genes and EA H1N1 internal genes (Campitelli et al. 1997; Castrucci et al. 1993). This H3N2 reassortant has been maintained in European countries since then and was introduced into Asian countries in the late 1990s (Gregory et al. 2001) (see below). The third major human H3N2 variant to cause zoonotic outbreaks in pigs was A/Sydney/05/97 (Syd). Introduction of this variant into pigs occurred, at least, in China and the USA (Peiris et al. 2001; Yu et al. 2008; Zhou et al. 1999). In China, the virus kept the entire human-like particle, while in the North America it further reassorted with CS and avian viruses to generate double and triple reassortant viruses (TR, H3N2, H1N2 and H1N1 subtypes) (Zhou et al. 1999; Karasin et al. 2006).

Since 2005, although H3N2 viruses have failed to be detected in the pig population under our surveillance program in Hong Kong (Vijaykrishna et al. 2011), contemporary and early human H3N2 variants were still isolated from pigs in the wider region (Hause et al. 2010; Lekcharoensuk et al. 2010; Kyriakis et al. 2011). Except for the major variants mentioned above, most human-like H3N2 variants seem to be transient and to have difficulty in becoming established in pigs as none have formed an independent group or sublineage in the evolutionary trees. Thus, it is likely that human H3N2 variants were regularly introduced into pigs, but most were prevalent at a low level within a small geographic location, and failed to become established.

2.3 European H3N2 Reassortant Viruses

In 1999, an H3N2 reassortant swine virus (represented by swine/Hong Kong/5212/99) was isolated in Hong Kong from pigs imported from southern China. This virus was antigenically and genetically distinct from the human-like H3N2 viruses then circulating in pigs. Genetic analyses revealed that this virus had PC-like surface genes and EA-like internal genes (Vijaykrishna et al. 2011). It was closely related phylogenetically to the European H3N2 reassortant viruses that were generated in the mid-1980s and have circulated in Europe since then (Claas et al. 1994; Campitelli et al. 1997; Castrucci et al. 1993). This virus caused a human infection case in Hong Kong (Gregory et al. 2001).

These findings provide a typical example of the direct introduction of swine influenza virus from Europe to Asia, very likely via pig movement. To largely increase its pig population and production, China started to import breeding pigs from European countries, such as Denmark (e.g. the DanBred organization) and set up breeding pig farms in the Zhujiang delta region in Guangdong since the mid-1990s. Unfortunately, the swine influenza virus was not included on Chinese warrant agent lists before 2009. This European H3N2 reassortant virus seemed to only circulate in Guangdong province in southern China and has not been reported from any other places in Asia.

2.4 European Avian-like H1N1

The first detection of EA viruses outside European countries occurred in early 2001 in Hong Kong. The virus (swine/Hong Kong/8512/01) (Smith et al. 2009; Vijaykrishna et al. 2011) was isolated from pigs imported from southern China. Since then, this virus has co-circulated with other swine influenza viruses, including CS, H3N2 and, later, American triple reassortant H1N2 (see below) viruses in this region and gradually became predominant from 2006. The replacement of CS viruses with EA viruses in pigs in the field took 4 to 5 years, similar to what happened in Europe after the EA lineage became established in pigs in the mid 1980s (Brown 2000). Phylogenetic analyses showed that the EA viruses isolated from pigs in China form a monophyletic group, suggesting a single introduction of this virus (Vijaykrishna et al. 2011).

H1N1 reassortants between CS and EA viruses were also detected in 2001 (Fig. 1), very likely occurring at the beginning of the introduction of EA viruses. Genetic analyses based on publicly available data showed that the EA virus was also introduced into pigs in Thailand around 2000 (Takemae et al. 2008). This virus also reassorted with CS viruses in pigs in this country (Takemae et al. 2008). Currently, EA viruses or its variants have become a major swine influenza lineage prevailing in the region. Whether the predominance of EA H1N1 viruses in the field is directly affecting the prevalence of H3N2 viruses is still unknown.

2.5 American Triple Reassortant Virus

In the reassortment event of 1998, both H3N2 and H1N2 triple reassortant viruses were generated in North America (Zhou et al. 1999; Karasin et al. 2006). Since 2002 American triple reassortant H1N2 viruses were regularly isolated from pigs in our surveillance program in China (Fig. 1). This virus has prevailed since then at generally low levels in pigs but was more prevalent during 2004 (Vijaykrishna et al. 2011). Our surveillance program suggests that triple reassortant viruses were introduced to China on several occasions (Vijaykrishna et al. 2011). From 2004, both H1N2 and H3N2 triple reassortant viruses have been isolated from pigs in Korea (Jung and Chae 2004; Pascua et al. 2008) Given that different subtypes with gene segments falling into polyphyletic groups were found in these two countries, separate introductions of the virus from the USA appear to have occurred.

2.6 Reassortant Viruses Between EA and TR

The 2009 pandemic H1N1 virus was derived by reassortment from several swine influenza viruses, which might include EA viruses (for the NA and M segments), European H3N2 reassortant viruses (for the M segment) and TR viruses (for the

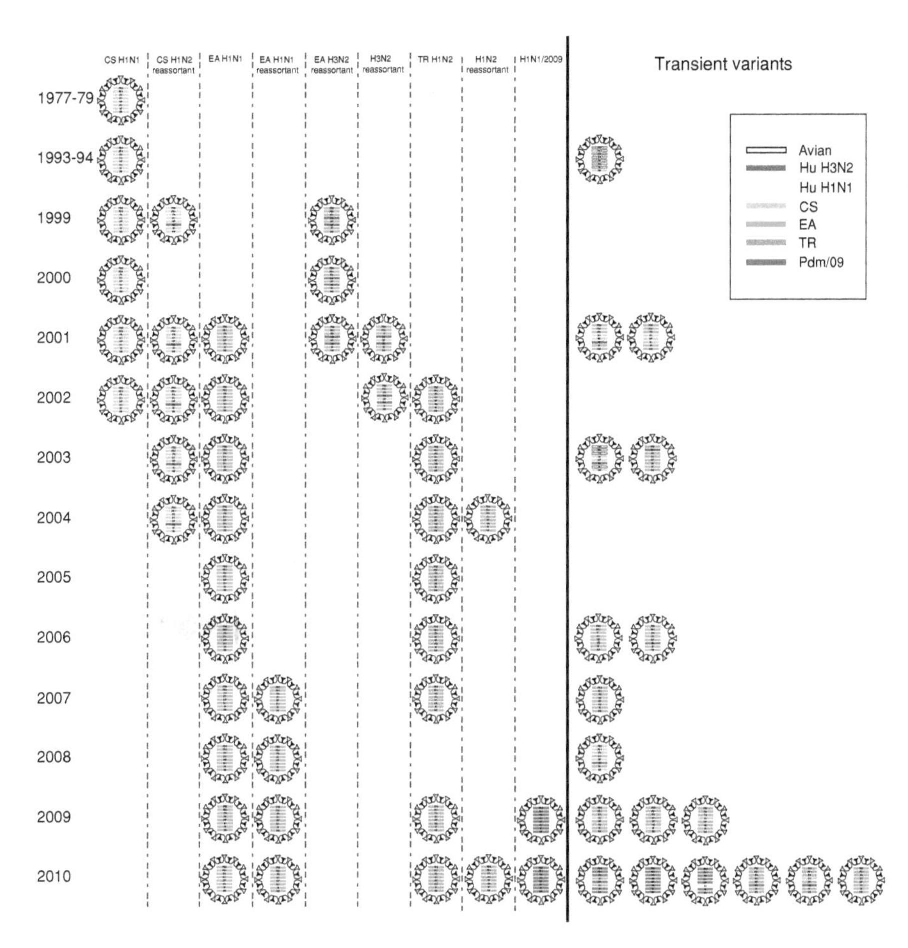

Fig. 1 Timeline of the genotype of viruses identified in surveillance program in China. Genotypes observed over at least two years are shown to the left. Segments are ordered by size from top to bottom for each genotype. The lineage of origin of each segment is indicated by *color*: *white* avian origin, *purple* human seasonal H3N2, *yellow* human seasonal H1N1, *blue* classical swine (CS), *green* European avian-like (EA), *orange* American triple reassortant (TR), *red* pandemic H1N1 2009 (Pdm/09)

remaining six segments) (Garten et al. 2009; Smith et al. 2009b). Thus, co-circulation of EA and TR viruses would appear to be essential for the genesis of this pandemic virus. Based on our surveillance findings and publicly available data, co-circulation of EA and TR viruses were observed in China from 2003 onwards (Fig. 1). However, EA-like viruses were never reported from America and TR-like viruses were also not reported from European countries. It is plausible, therefore, that the 2009 pandemic H1N1 virus could have been generated within the pigs of an Asian country (most likely China). Indeed, reassortment events between EA and TR viruses are not rare, and reassortant swine viruses with differing genotypes have been recognized in the field (Fig. 1). Reassortants between EA and TR viruses, which are the most closely related to the 2009 H1N1 pandemic virus (by having seven gene segments from the same lineages), were isolated in southern China on four sampling occasions (two before and two after the pandemic). However, all these reassortants had different evolutionary pathways, and almost all reassortants between EA and TR viruses appear to be transient. The only exception was a reassortant which had seven segments from the EA lineage and the NS segment from the TR lineage and has become established in the field and may eventually predominant in pigs in China (Vijaykrishna et al. 2011) (Fig. 1).

2.7 *Pandemic/2009-like H1N1 and its Variants*

From the time of the peak of the human pandemic, the pandemic H1N1/09-like virus has been repeatedly isolated from pigs in many Asian countries (Vijaykrishna et al. 2010; Song et al. 2010). Most of these pdm/09-like swine isolates resulted from different direct introductions from humans to pigs. However, the detection of multiple reassortant viruses between pdm/09 and other swine viruses (Vijaykrishna et al. 2010; Starick et al. 2011; Moreno et al 2010), along with the high seroconversion rate to the pandemic virus in pigs (unpublished data), suggest that this virus might gradually become established in this host. Questions remaining to be answered are when that will occur, what kind of genetic composition the established virus will have and what its long-term impact will be.

2.8 *Avian-like Influenza Viruses*

In the last two decades avian influenza viruses have frequently been isolated or detected from pigs in Asian countries, likely due to increased farming activity and interaction between pigs and birds. However, none of these interspecies transmission events caused severe consequences or the establishment of avian origin viruses or virus genes in pigs. The most frequently detected avian viruses in pigs in Asian countries belong to the virus lineages that are long-term enzootic in poultry, such as H9N2 and H5N1 viruses. H9N2 avian-like viruses were detected in China and Korea, and H5N1 avian-like viruses were reported from China, Vietnam and Indonesia (Nguyen et al. 2005; Nidom et al. 2010; Yu et al. 2011). All H9N2 and H5N1 swine virus isolates were from different sublineages and variants, highlighting the long-term potential threat from these viruses.

Other subtypes of avian influenza viruses detected in pigs in Asian countries include H1N1, H3N2, H5N2, H11N6 and H6N6 (Guan et al. 1996; Zhang et al. 2011; Lee et al. 2009; Kim et al. 2010; Kida et al. 1988). Genetic analyses revealed that these viruses were likely derived from those residential in aquatic birds. Almost all were detected only on a single sampling occasion, but some were from disease surveillance in pigs and low seroconversion rates were observed (Zhang et al. 2011; Lee et al. 2009). This shows that pigs are susceptible to most subtypes of avian influenza viruses. A major concern is that should highly pathogenic H5N1 Asian lineage viruses recruit mammalian-adapted virus genes from pigs, a human-to-human transmissible virus might be generated. Systematic surveillance in different countries would be greatly helpful to counteract such an event.

3 Summary

The findings presented here suggest that almost all major swine influenza virus lineages from different continents are co-circulating in pigs in Asian countries. The movement of live pigs between different continents is probably responsible for this. Co-circulation of these different virus lineages will naturally increase virus interaction and reassortment, and the genetic diversity in swine influenza viruses. The emergence of the 2009 pandemic H1N1 virus provided the clearest example of the potential consequences of co-circulating viruses even though we lack evidence to show that this pandemic virus was initially generated in Asian countries. Given that all current swine virus lineages and their constituent segments have been prevalent for more than a decade in pigs (i.e. they are fully mammalian-adapted) and H9N2 and H5N1 avian influenza viruses are widely enzootic in poultry in the region, generation of a novel virus with efficient transmissibility in pigs or even in humans is possible.

Acknowledgments This work was supported by the National Institutes of Health (National Institute of Allergy and Infectious Diseases contract HSN266200700005C); Li Ka Shing Foundation; and Area of Excellence Scheme of the University Grants Committee of the Hong Kong SAR (grant AoE/M-12/06).

References

Arikawa J, Yamane N, Odagiri T, Ishida N (1979) Serological evidence of H1 influenza virus infection among Japanese hogs. Acta Virol 23:508–511

Arikawa J, Yamane N, Totsukawa K, Ishida N (1982) The follow-up study of swine and Hong Kong influenza virus infection among Japanese hogs. Tohoku J Exp Med 136:353–358

Brown IH (2000) The epidemiology and evolution of influenza viruses in pigs. Vet Microbiol 74:29–46. doi:S0378-1135(00)00164-4[pii]

Campitelli L et al (1997) Continued evolution of H1N1 and H3N2 influenza viruses in pigs in Italy. Virology 232 :310–318. doi:S0042-6822(97)98514-7[pii]10.1006/viro.1997.8514

Castrucci MR et al (1993) Genetic reassortment between avian and human influenza A viruses in Italian pigs. Virology 193:503–506. doi:S0042-6822(83)71155-4[pii]10.1006/viro.1993.1155

Chatterjee S, Mukherjee KK, Mondal MC, Chakravarti SK, Chakraborty MS (1995) A serological survey of influenza a antibody in human and pig sera in Calcutta. Folia Microbiol (Praha) 40:345–348

Chun JWH (1919) Influenza, including its infection among pigs. Natl Med J China 5:34–44

Claas EC, Kawaoka Y, de Jong JC, Masurel N, Webster RG (1994) Infection of children with avian–human reassortant influenza virus from pigs in Europe. Virology 204:453–457. doi: S0042-6822(84)71553-4[pii]10.1006/viro.1994.1553

Claas EC et al (1998) Human influenza A H5N1 virus related to a highly pathogenic avian influenza virus. Lancet 351:472–477. doi:S0140-6736(97)11212-0[pii]10.1016/S0140-6736(97)11212-0

Das KP, Mallick BB, Das K (1981) Note on the prevalence of influenza antibodies in swine. Ind J Anim Sci 51:907–908

Dawood FS et al (2009) Emergence of a novel swine-origin influenza A (H1N1) virus in humans. N Engl J Med 360:2605–2615. doi:NEJMoa0903810 [pii] 10.1056/NEJMoa0903810

Garten RJ et al (2009) Antigenic, genetic characteristics of swine-origin 2009 A(H1N1) influenza viruses circulating in humans. Science 325:197–201. doi:1176225[pii]10.1126/science.1176225

Gorman OT et al (1991) Evolution of influenza A virus nucleoprotein genes: implications for the origins of H1N1 human and classical swine viruses. J Virol 65:3704–3714

Gregory V et al (2001) Infection of a child in Hong Kong by an influenza A H3N2 virus closely related to viruses circulating in European pigs. J Gen Virol 82:1397–1406

Guan Y et al (1996) Emergence of avian H1N1 influenza viruses in pigs in China. J Virol 70: 8041–8046

Hause BM, Oleson TA, Bey RF, Stine DL, Simonson RR (2010) Antigenic categorization of contemporary H3N2 Swine influenza virus isolates using a high-throughput serum neutralization assay. J Vet Diagn Invest 22:352–359. doi:22/3/352 [pii]

Jung K, Chae C (2004) Phylogenetic analysis of an H1N2 influenza A virus isolated from a pig in Korea. Brief Report. Arch Virol 149:1415–1422. doi:10.1007/s00705-004-0324-9

Jung K, Song DS (2007) Evidence of the co-circulation of influenza H1N1, H1N2 and H3N2 viruses in the pig population of Korea. Vet Rec 161:104–105. doi:161/3/104[pii]

Kanegae Y, Sugita S, Shortridge KF, Yoshioka Y, Nerome K (1994) Origin and evolutionary pathways of the H1 hemagglutinin gene of avian, swine and human influenza viruses: cocirculation of two distinct lineages of swine virus. Arch Virol 134:17–28

Karasin AI, Carman S, Olsen CW (2006) Identification of human H1N2, human–swine reassortant H1N2, H1N1 influenza A viruses among pigs in Ontario, Canada (2003 to 2005). J Clin Microbiol 44:1123–1126. doi:44/3/1123[pii]10.1128/JCM.44.3.1123-1126.2006

Kida H, Shortridge KF, Webster RG (1988) Origin of the hemagglutinin gene of H3N2 influenza viruses from pigs in China. Virology 162:160–166

Kilbourne ED (2006) Influenza pandemics of the 20th century. Emerg Infect Dis 12:9–14

Kim HR et al (2010) Characterization of H5N2 influenza viruses isolated in South Korea and their influence on the emergence of a novel H9N2 influenza virus. J Gen Virol 91:1978–1983. doi:vir.0.021238-0 [pii]10.1099/vir.0.021238-0

Koen JS (1919) A practical method for field diagnoses of swine diseases. Am J Vet Med 14:468–470

Kundin WD (1970) Hong Kong A-2 influenza virus infection among swine during a human epidemic in Taiwan. Nature 228:857

Kupradinun S et al (1991) The first isolation of swine H1N1 influenza viruses from pigs in Thailand. Arch Virol 118:289–297

Kyriakis CS et al (2011) Virological surveillance, preliminary antigenic characterization of influenza viruses in pigs in five European countries from 2006 to 2008. Zoonoses Public Health 58:93–101. doi:10.1111/j.1863-2378.2009.01301.xJVB1301[pii]

Lee CS et al (2008) Phylogenetic analysis of swine influenza viruses recently isolated in Korea. Virus Genes 37:168–176. doi:10.1007/s11262-008-0251-z

Lee JH et al (2009) Isolation and genetic characterization of H5N2 influenza viruses from pigs in Korea. J Virol 83 :4205–4215. doi:10.1128/JVI.02403-09

Lekcharoensuk P, Nanakorn J, Wajjwalku W, Webby R & Chumsing W (2010) First whole genome characterization of swine influenza virus subtype H3N2 in Thailand. *Vet Microbiol* 145 :230–244. doi:S0378-1135(10)00189-6[pii]10.1016/j.vetmic.2010.04.008

Lin YP et al (2000) Avian-to-human transmission of H9N2 subtype influenza A viruses: relationship between H9N2 and H5N1 human isolates. Proc Natl Acad Sci U S A 97:9654–9658. doi:10.1073/pnas.160270697160270697[pii]

Mohan R, Saif YM, Erickson GA, Gustafson GA, Easterday BC (1981) Serologic and epidemiologic evidence of infection in turkeys with an agent related to the swine influenza virus. Avian Dis 25:11–16

Moreno A et al (2010) Novel H1N2 swine influenza reassortant strain in pigs derived from the pandemic H1N1/2009 virus. Vet Microbiol. doi:S0378-1135(10)00584-5[pii]10.1016/j.vetmic. 2010.12.011

Nerome K et al (1981) Antigenic and genetic analysis of A/Hong Kong (H3N2) influenza viruses isolated from swine and man. J Gen Virol 56:441–445

Nerome K, Ishida M, Oya A, Oda K (1982) The possible origin H1N1 (Hsw1N1) virus in the swine population of Japan and antigenic analysis of the isolates. J Gen Virol 62(Pt 1):171–175

Nguyen DC et al (2005) Isolation and characterization of avian influenza viruses, including highly pathogenic H5N1, from poultry in live bird markets in Hanoi, Vietnam, in 2001. J Virol 79:4201–4212. doi:79/7/4201[pii]10.1128/JVI.79.7.4201-4212.2005

Nidom CA et al (2010) Influenza A (H5N1) viruses from pigs, Indonesia. Emerg Infect Dis 16:1515–1523

Ogawa Y et al (1983) Sero-epizootiological study on swine influenza in a prefecture of Japan. J Hyg (Lond) 90:403–406

Ouchi A et al (1996) Large outbreak of swine influenza in southern Japan caused by reassortant (H1N2) influenza viruses: its epizootic background and characterization of the causative viruses. J Gen Virol 77(Pt 8):1751–1759

Pascua PN et al (2008) Seroprevalence and genetic evolutions of swine influenza viruses under vaccination pressure in Korean swine herds. Virus Res 138 :43–49. doi:S0168-1702(08) 00292-X[pii]10.1016/j.virusres.2008.08.005

Peiris M et al (1999) Human infection with influenza H9N2. Lancet 354:916–917. doi:S0140673699033115[pii]

Peiris JS et al (2001) Cocirculation of avian H9N2 and contemporary "human" H3N2 influenza A viruses in pigs in southeastern China: Potential for genetic reassortment? J Virol 75: 9679–9686, doi:10.1128/JVI.75.20.9679-9686.2001

Pensaert M, Ottis K, Vandeputte J, Kaplan MM, Bachmann PA (1981) Evidence for the natural transmission of influenza A virus from wild ducts to swine and its potential importance for man. Bull World Health Organ 59:75–78

Pereda A et al (2010) Pandemic (H1N1) 2009 outbreak on pig farm, Argentina. Emerg Infect Dis 16:304–307

Reid AH, Taubenberger JK (2003) The origin of the 1918 pandemic influenza virus: a continuing enigma. J Gen Virol 84:2285–2292

Scholtissek C (1990) Pigs as the "mixing vessel" for the creation of new pandemic influenza A viruses. Med Princip Prac 2:65–71

Scholtissek C, Burger H, Kistner O, Shortridge KF (1985) The nucleoprotein as a possible major factor in determining host specificity of influenza H3N2 viruses. Virology 147:287–294

Shope RE (1931) Swine Influenza : Iii. Filtration Experiments and Etiology. J Exp Med 54: 373–385

Shortridge KF, Webster RG (1979) Geographical distribution of swine (Hsw1N1) and Hong Kong (H3N2) influenza virus variants in pigs in Southeast Asia. Intervirology 11:9–15

Shortridge KF, Webster RG, Butterfield WK, Campbell CH (1977) Persistence of Hong Kong influenza virus variants in pigs. Science 196:1454–1455

Shortridge KF, Cherry A, Kendal AP (1979) Further studies of the antigenic properties of H3N2 strains of influenza A isolated from swine in South East Asia. J Gen Virol 44:251–254

Smith GJ et al (2009a) Dating the emergence of pandemic influenza viruses. Proc Natl Acad Sci U S A 106:1709–11712. doi:0904991106[pii]10.1073/pnas.0904991106
Smith GJ et al (2009b) Origins, evolutionary genomics of the 2009 swine-origin H1N1 influenza A epidemic. Nature 459:1122–1125. doi:nature08182[pii]10.1038/nature08182
Song DS et al (2007) Isolation and phylogenetic analysis of H1N1 swine influenza virus isolated in Korea. Virus Res 125:98–103. doi:S0168-1702(06)00364-9[pii]10.1016/j.virusres.2006.11.008
Song DS et al (2003) Isolation of H3N2 swine influenza virus in South Korea. J Vet Diagn Invest 15:30–34
Song MS et al (2010) Evidence of human-to-swine transmission of the pandemic (H1N1) 2009 influenza virus in South Korea. J Clin Microbiol 48:3204–3211. doi:JCM.00053-10[pii]10.1128/JCM.00053-10
Starick E et al (2011) Re-assorted pandemic (H1N1) 2009 influenza A virus discovered from pigs in Germany. J Gen Virol. doi:vir.0.028662-0[pii]10.1099/vir.0.028662-0
Subbarao K et al (1998) Characterization of an avian influenza A (H5N1) virus isolated from a child with a fatal respiratory illness. Science 279:393–396
Sugimura T, Yonemochi H, Ogawa T, Tanaka Y, Kumagai T (1980) Isolation of a recombinant influenza virus (Hsw 1 N2) from swine in Japan. Arch Virol 66:271–274
Takemae N et al (2008) Genetic diversity of swine influenza viruses isolated from pigs during 2000 to 2005 in Thailand. Influenza Other Respi Viruses 2:181–189. doi:IRV062[pii]10.1111/j.1750-2659.2008.00062.x
Vijaykrishna D et al (2011) Long-term evolution and transmission dynamics of swine influenza A virus. Nature 473:519–522
Vijaykrishna D et al (2010) Reassortment of pandemic H1N1/2009 influenza A virus in swine. Science 328:1529. doi:328/5985/1529[pii]10.1126/science.1189132
Webster RG, Hinshaw VS, Bean W J Jr, Turner B, Shortridge KF (1977) Influenza viruses from avian and porcine sources and their possible role in the origin of human pandemic strains. Dev Biol Stand 39:461–468
Webster RG, Peiris M, Chen H, Guan Y (2006) H5N1 outbreaks and enzootic influenza. Emerg Infect Dis 12:3–8
Weingartl HM et al (2010) Genetic, pathobiologic characterization of pandemic H1N1 2009 influenza viruses from a naturally infected swine herd. J Virol 84:2245–2256. doi:JVI.02118-09[pii]10.1128/JVI.02118-09
Yamane N, Arikawa J, Odagiri T, Kumasaka M, Ishida N (1978) Distribution of antibodies against swine and Hong Kong influenza viruses among pigs in 1977. Tohoku J Exp Med 126:199–200
Yamane N, Arikawa J, Odagiri T, Ishida N (1979) Annual examination of influenza virus infection among pigs in Miyagi prefecture, Japan: the appearance of Hsw1N1 virus. Acta Virol 23:240–248
Yip TKS (1976) Serological survey on the influenza antibody status in pigs of the Takwuling pig breeding centre. Agric Hong Kong 1:446–458
Yu H et al (2011) Genetic diversity of H9N2 influenza viruses from pigs in China: a potential threat to human health? Vet Microbiol 149:254–261. doi:S0378-1135(10)00520-1[pii]10.1016/j.vetmic.2010.11.008
Yu H et al (2008) Genetic evolution of swine influenza A (H3N2) viruses in China from 1970 to 2006. J Clin Microbiol 46:1067–1075. doi:JCM.01257-07[pii]10.1128/JCM.01257-07
Zhang G et al (2011) Identification of an H6N6 swine influenza virus in southern China. Infect Genet Evol. doi:S1567-1348(11)00065-7[pii]10.1016/j.meegid.2011.02.023
Zhou NN et al (1999) Genetic reassortment of avian, swine, and human influenza A viruses in American pigs. J Virol 73:8851–8856

Clinicopathological Features of Swine Influenza

B. H. Janke

Abstract In this chapter, the clinical presentations, the development of infection and the macroscopic and microscopic lesions of swine influenza virus (SIV) infection are described. Both natural and experimental infections are discussed.

Contents

B. H. Janke (✉)
Veterinary Diagnostic Laboratory, Department of Veterinary Diagnostic and Production Animal Medicine, College of Veterinary Medicine, Iowa State University, Ames, IA 50011, USA
e-mail: bhjanke@iastate.edu

Current Topics in Microbiology and Immunology (2013) 370: 69–83
DOI: 10.1007/82_2013_308

Published Online: 13 February 2013

1 Introduction

As the worst global human pandemic climaxed during the summer of 1918, a new disease entity began to be recognized in swine in the Midwestern United States. The clinical signs of this new disease readily differentiated it from classical swine fever (hog cholera), which was the infectious disease of most consequence for swine at that time. This new respiratory disease was tabbed "hog flu" because of the similarity of the disease to influenza in humans (Dorset et al. 1922).

An early description paints a memorable clinical picture: "The onset of hog flu, as already stated, is sudden, an entire herd coming down, as a rule, within a day or two…The first symptom noted is loss of appetite, the animals failing to come up for their feed. They are disinclined to move and lie around the straw stacks or in their houses. When temperatures are taken, the animals are found to have fever. A thumpy or jerky respiration soon develops which is best observed when the animals are lying down and at first may be so slight as to escape notice unless the animals are carefully watched; later, it becomes more pronounced and may be noted when animals are standing. The disease evidently has a very short incubation period and develops rapidly. The second or third day, the entire herd, as a rule, will be lying in their nests and often present a very sick appearance. Sometimes one may walk among the sick animals and even step over them without rousing them, and anyone viewing for the first time a herd suffering from hog flu at the height of the infection would probably think that most of the affected animals would succumb. When the sick animals are roused from their nests, they almost invariably cough. The cough is paroxysmal in character, the back being often arched, and the spells of coughing are sometimes of sufficient violence to induce vomiting; in this respect the disease resembles whooping-cough in the human. When the paroxysms of coughing have passed, the animals stand in a listless attitude with their heads down, their tails limp, and soon lie down as though tired. The sick animals usually rest on their bellies, and sometimes assume a partly sitting position with the body propped on the forelegs, as if to afford room for greater lung expansion. There is usually a conjunctivitis, characterized by a watery or gummy secretion from the eyes, and a nasal discharge may also be present……" (McBryde 1927).

2 Clinical Disease

2.1 Classical Epidemic Swine Influenza

Epidemic swine influenza as described above was the predominant presentation of the disease in the United States for nearly 70 years. Indeed, this classical virus and presentation of swine influenza in conventional swine populations continues today as an acute, high morbidity-low mortality infection that spreads rapidly through groups of pigs. For the first 1–2 days after infection, affected pigs develop high fevers (>105°F, 40.5 °C) with lethargy and anorexia. Close examination reveals clear nasal discharge and conjunctivitis. Tachypnea and expiratory dyspnea (thumping) are often pronounced, especially when pigs are forced to move. By days 3 and 4, pigs begin to acquire the hallmark clinical sign of this disease, a harsh deep barking cough that results from the extensive bronchitis and bronchiolitis. In many pigs, fevers will have begun to drop by this time. Pigs of all ages, from mature sows to nursery pigs, may be similarly affected, but clinical disease is often milder in nursing pigs. In some outbreaks, sows may be inappetent and lethargic and develop high fevers but have less prominent respiratory clinical signs. Pregnant animals may abort. In the absence of concurrent bacterial pneumonia, individual pigs recover quickly, usually within 6–7 days. Not all pigs in a group will be infected simultaneously, and the disease course for the entire group may require a 2+ weeks before clinical signs abate and pigs return to normal body condition and weight gain. Mortality is generally low although some virus strains exact a higher toll.

Historically, there has always been a distinct seasonality to swine influenza in the north central U. S., a part of the country that has widely divergent seasonal temperatures. In years past, outbreaks commonly occurred each fall, with little recognition of the disease at other times of year. Even under current confinement production systems, the disease still exhibits a consistent seasonality, with the greatest peak in the fall and a smaller peak in the spring, the seasons of transition for prevailing weather conditions (Janke et al. 2000). Wide swings in temperature over short periods of time make it difficult to modulate housing environments, and these climatic stresses may increase animal susceptibility to infection. Cool moist conditions contribute to environmental survival and aerosol spread of the virus. How and where the virus survived/survives between outbreaks has never been fully explained. A long-term carrier state has not been discovered, and most individual pigs appear to clear the virus within 2 weeks. The virus is likely maintained in herds by subclinical passage to naïve pigs or those with low or compromised immunity. Studies on vaccine efficacy have indicated that there is no absolute immunity threshold for swine influenza virus (SIV) infection. Pigs with sufficient immunity to prevent clinical illness can still be infected and shed virus, though it is much reduced in duration and titer (Richt et al. 2006; Van Reeth et al. 2001).

2.2 *Current Clinical Expression in Larger Swine Populations in Segregated Rearing Production Systems*

Many pigs today are raised in segregated rearing systems. Sows are maintained, bred, and farrowed at a location separate from the farms on which younger pigs are fed to market weight. Once farrowed, pigs are only kept with the sows for about 3 weeks before they are weaned and moved to nurseries on another site (segregated early weaning). In two-site systems, pigs will be raised at that location until being sent to market. In three-site systems, pigs will be fed at the nursery site until about 10–12 weeks of age, after which they are moved to grow-finish units at a third location. At each location, the goal is to fill and empty each building completely at one time (all-in all-out). The objective is disease control, i.e., to minimize situations in which infections can be passed from older immune pigs on a site to younger naïve pigs as they are brought into the same buildings. In continuous flow systems, viruses and bacteria have greater opportunity to maintain contagious levels because of the periodic addition of susceptible hosts. This age-based segregation and movement of pigs must be considered when considering the clinical presentation of swine influenza as it currently manifests itself. In nearly all swine production units, one can find influenza virus in circulation or serologic evidence of previous exposure.

At the present time, because of the almost universal immunity against SIV at some level against some variant in all herds, and the number of subtypes, antigenic clusters within subtypes and reassortants that circulate within swine populations, expression varies widely. Outbreaks still occur but infections are more frequently endemic with clinical expression more muted and melded with that of other concurrent respiratory infections. Pigs of all ages from nursing pigs to sows may be infected, with blips of clinical disease appearing at different phases of production that vary with the situation. It becomes the task of veterinary practitioners and diagnosticians to determine whether increased clinical disease is due to the resurgence of a virus already circulating within the operation or to the introduction of a different virus against which only partial immunity may exist.

Expression of disease generally falls into one of the following presentations:

1. Acute, "fulminating" SIV—Resembling disease as it was first observed 90 years ago but now the least frequent presentation, this is the outbreak of a pathogenic variant in a population that has little or no specific antibody to attenuate the infection. There is very rapid spread, high fevers, anorexia, expiratory dyspnea with effort, mortality, and a brief course in the herd/group. Mortality occurs in pigs whose lungs fill rapidly with fluid; a typical 'foam cast' forms in the bronchial tree and is expelled through the nose. There is very rapid spread within and between sites and through pigs of all ages. Even nursing pigs are severely affected, but essentially all survive. Recovery is rapid and often complete, with very few secondarily infected or chronically affected pigs, and infection is followed by very high levels of specific immunity. This is rarely confounded with other agents even though they may be present.
2. Age-associated influenza in growing pigs—Infection can be relatively predictable in certain systems depending on how they are structured. Maternal passively acquired antibody is protective for weaned pigs against endemic virus, but that immunity wanes with maternal antibody decay. Usually occurring when sow herd immunity is uniform, and pigs have roughly equivalent levels of passive protection, infection that is clinically evident is delayed until the pigs are exposed to homologous or variant virus that can 'break through' the collective passive immunity. The classic pattern is robust pigs with no problems through the nursery phase until they are 10–12 weeks old and in finishers where

they are exposed to other older growing pigs that are shedding virus. Approximately 2 weeks into the finishing phase, pigs develop cough and depression that does not explode, but rather works its way through the group over a protracted 2–3 week timeline as the individual passive antibody decay curves meet up with the various loads of virus that overcome it. The actual SIV-induced clinical disease expressed is usually not dramatic, of variable severity and commonly complicated by concurrent infections endemic in the group.
3. Piglet influenza—A consistent/persistent clinical expression of SIV can be found in nursing to newly weaned pigs. Again associated with passive protection variance within groups and increasing frequency of SIV variants, clinically the infection is first appreciated as a cough in the 2–3-week-old pig about to be weaned. But the cough is in scattered pigs in a given room (maybe 10 % of litters have a pig that sporadically coughs) that are hard to find and almost too subtle to raise any alarm. When these pigs are moved to the nursery, the stress and activity exacerbate clinical expression. During the move and on first entry into the nursery, there is a common description "…newly weaned pigs came off the truck coughing…" even though the farrowing management would say "…no, we didn't notice any cough; maybe a pig or two…". Clinically in this situation, a subset of pigs suffers fever, anorexia, and after a couple days, cough due to SIV infection. Again, the severity of clinical disease is greatly dependent on the level and specificity of passive protection. Affected pigs are anorexic during the critical transition phase to solid food, i.e., they are sick for 36 to 48 h after the move, and become very hard to start on feed. When just a few pigs in a large group are affected, the infection probably goes unrecognized, but on many farms, a consistent 5–10 % or greater are affected and the problem is economically substantial. These situations are the purpose of sow immunization, not to protect the sow, but to try to extend the passive shield until pigs are well-started on feed and can handle the infection.

2.3 *Clinical Disease in Experimentally Infected Pigs*

The clinical disease induced by experimental challenge rarely reaches the severity of that observed in the field (Landolt et al. 2003; De Vleeschauwer et al. 2009). Clinical signs of illness (fever, depression, anorexia, tachypnea, serous nasal, and ocular discharge) develop to some degree after most challenges but vary under the multitude of protocols employed (Nayak et al. 1965; Winkler and Cheville 1986; Brown et al. 1993; Van Reeth et al. 1996; Thacker et al. 2001; Richt et al. 2003; Landolt et al. 2003; Jung et al. 2005; Vincent et al. 2006; De Vleeschauwer et al. 2009; Sreta et al. 2009). Under most experimental models, tachypnea is evident in pigs at least when aroused, but coughing is usually minimal, limited to an occasional soft cough that develops a day or two after the onset of other clinical signs. Fever is quite variable as is the onset of clinical signs. In general, with higher virus titers in inoculum (>10^6–10^7 $TCID_{50}$, EID_{50} or PFU/pig) and intratracheal inoculation, the onset of illness occurs sooner [24–36 h postinoculation (PI)] and clinical signs are more apparent. With lesser challenge regimens (10^3–10^5 $TCID_{50}$, EID_{50}, PFU/pig) and intranasal inoculation, clinical disease may not be seen or may take 2–4 days to become evident. Duration of clinical illness is usually 2–4 days with most experimental infections.

3 Virus Infection, Replication, and Shedding

SIVs are spread between pigs through direct contact via nasal secretions and through inhalation of aerosolized virus in droplets generated by coughing and exhalation. The cellular targets of infection are the epithelial cells lining the nasal passages, trachea, bronchi, bronchioles, and alveoli. Numerous experimental infection studies have been conducted to define the progression of the infectious process.

A variety of inoculation methods have been employed in these experimental infection studies to administer virus: nebulization via nose cone or chamber, direct intranasal (IN) inoculation, and intratracheal (IT) inoculation. Nebulization is probably the most effective method of depositing large quantities of virus throughout the respiratory tract but this method is also the most labor intensive and is little used. Intranasal methods (the mechanical specifics of which are often not described) appear to be more variable in their efficacy of establishing infection, probably because with some techniques pigs swallow most of the challenge dose. Intratracheal methods provide the most consistent challenges, and if not deposited too far down the tract, the virus appears to be well-distributed throughout the lungs.

Infection and multiplication in host cells progresses very rapidly with influenza viruses. In an immunofluorescent study (Nayak et al. 1965), the first evidence of virus infection was a pale fluorescence in the nucleus of bronchial epithelial cells, as early as 2 h PI. By 4 h PI, virus antigen was abundant in both nucleus and cytoplasm of infected cells. In an ultrastructural study (Winkler and Cheville 1986), virus was observed budding from the surface of Type II pneumocytes as early as 5 h PI.

Only low numbers of randomly scattered cells are observed in the nasal turbinates and trachea during the first 24–72 h PI (Nayak et al. 1965). The most extensive infection occurs in epithelial cells lining the bronchi and bronchioles, with peak infection occurring at 48–72 h PI. Although some virus reaches the alveolar level early, especially with nebulization or high dosages with other methods, more extensive spread of virus to alveolar epithelial cells tends to occurs later in the course of infection, at 72–96 h PI (Jung et al. 2005; Van Reeth and Pensaert 1994).

Multiple studies (Brown et al. 1993; Van Reeth et al. 1996; Landolt et al. 2003; Vincent et al. 2006, 2009a, b; De Vleeschauwer et al. 2009; Ma et al. 2010) report nasal virus shedding by 1–3 days PI, regardless of the route of inoculation, and the duration of shedding, for 4–5 days, occasionally to 7 days PI. In one study, intranasal inoculation (10^7 EID_{50}) resulted in virus shedding by 24 h PI, whereas intratracheal inoculation resulted in nasal shedding being delayed until 72 h PI and the peak titer of virus shed was much lower (De Vleeschauwer et al. 2009). In another study using intranasal inoculation, lower virus titers (10^3–10^4 $TCID_{50}$) resulted in 24 h delay in onset and peak of nasal shedding compared to pigs given 10^5–10^6 $TCID_{50}$ although peak titers of shed virus were similar (10^7 $TCID_{50}$/ml) (Landolt et al. 2003). The amount of virus shed in nasal secretions tended to be fairly consistent through days 2–4 PI with peak titers described in the range of $10^{3.5}$–$10^{7.5}$ $TCID_{50}$, EID_{50}, or PFU/ml of fluid used to flush nasal passages or to wash virus from nasal swabs.

Determination of virus titers in lung homogenate or bronchoalveolar lavage fluids is often used in experimental studies to monitor the dynamics of virus production in the lung. Peak virus load in the lung, as measured in studies that cover the first few days of infection, occurs at about day 3 PI with titers varying from $10^{4.5}$ to $10^{8.3}$ $TCID_{50}$, EID_{50} or PFU/ml. Titers hold at relatively similar levels through 5 days PI. (Van Reeth et al. 1996; De Vleeschauwer et al. 2009; Vincent et al. 2009a, b; Ma et al. 2010).

The narrow time frame of virus replication and shedding described above is consistent in most challenge trials for most influenza viruses isolated from swine regardless of subtype. Experimentally, the dose of virus that reaches the lung initially may affect the course and severity of experimental infection. If low doses are given, either intranasally or intratracheally, the virus may initially spread more slowly which may result in a delayed onset and ultimately milder course of clinical disease. With most viruses, the course of infection is short and essentially complete within 5–7 days. The comparative effects of virus titer in inoculum, route of inoculation and/or age on the dynamics of infection have been described in several studies (Landolt et al. 2003; Richt et al. 2003; De Vleeschauwer et al. 2009).

In swine, influenza virus infection is generally considered to be limited to the respiratory tract, but a few studies have reported virus in extra-respiratory sites. A few infected cells were detected by immunohistochemistry in mediastinal lymph nodes, but none were detected in tonsil (Nayak et al. 1965). Influenza virus was isolated from the serum of all five inoculated pigs, for only one day each, at 1–3 days PI (Brown et al. 1993). In a more recent study, virus was detected by RT-PCR in spleen, ileum, and colon but not by virus isolation. Virus was detected in brainstem by both PCR assay and virus isolation, but no specific infected cells were detected by IHC (De Vleeschauwer et al. 2009). In this paper, researchers referred to unpublished data from in vitro studies indicating this virus could infect porcine trigeminal ganglion via the axons.

4 Pathology

4.1 Macroscopic Lesions

The most common macroscopic manifestation of influenza virus infection is a cranioventral bronchopneumonia that can affect a variable amount of the lung. The lesions are similar in both experimental inoculation and uncomplicated natural infections (Janke 1998) (Fig. 1a, b). Such expression would be expected since the virus enters the lung via the airways rather than through viremia. In milder infections, dark red multilobular to coalescing, often somewhat linear, foci of consolidation are evident in the hilar area and more dorsal portions of the cranial and middle lung lobes. More extensive infections involve larger, usually more ventral, portions of the cranial and middle lobes and cranioventral portions of the caudal lobe; as much as 40 % of the total lung volume may be affected. In field cases, the lesions often involve concurrent bacterial bronchopneumonia which results in more extensive lesions. In an occasional pig, a few hemorrhagic emphysematous bullae distending interlobular spaces may be evident. Tracheobronchial lymph nodes are variably swollen and congested. The trachea and nasal turbinates may be congested but are usually unremarkable. Although virus infects the epithelial lining of these upper airways, grossly visible necrosis does not develop.

Less frequently encountered in field situations, and not reproduced by experimental challenge, severe acute influenza infections may result in a diffusely congested and edematous lung with abundant foam in the trachea and larger airways (Janke 1998). In such an acutely affected lung, cranioventral lobular consolidation may be obscured by the diffuse inflammation.

In experimental studies (Winkler and Cheville 1986; Van Reeth et al. 1996; Thacker et al. 2001; Richt et al. 2003; Landolt et al. 2003; Vincent et al. 2006, 2009a, b; De Vleeschauwer et al. 2009; Sreta et al. 2009; Ma et al. 2010) the extent of lung involvement also is quite variable and usually is expressed as percent of total lung affected, calculated either by addition of the portions of each lobe (Halbur et al. 1995) or as the average for all lobes, with described values ranging from <1 to 58 %. As with clinical signs and virus distribution and shedding, lesion severity is influenced by the route of inoculation and the virus titer in the inoculum. An inoculum containing 10^6 or higher infectious doses of virus introduced intratracheally will often result in 10–30 % lung involvement. Inocula with 10^3–10^4 $TCID_{50}$ or EID_{50} of virus, especially if administered intranasally, may result in <10 % lung involvement. In some experimental infections, the dorsocaudal aspect of the caudal lobe also is affected, most likely an artifact of the method of inoculation as this presentation is unusual in field cases.

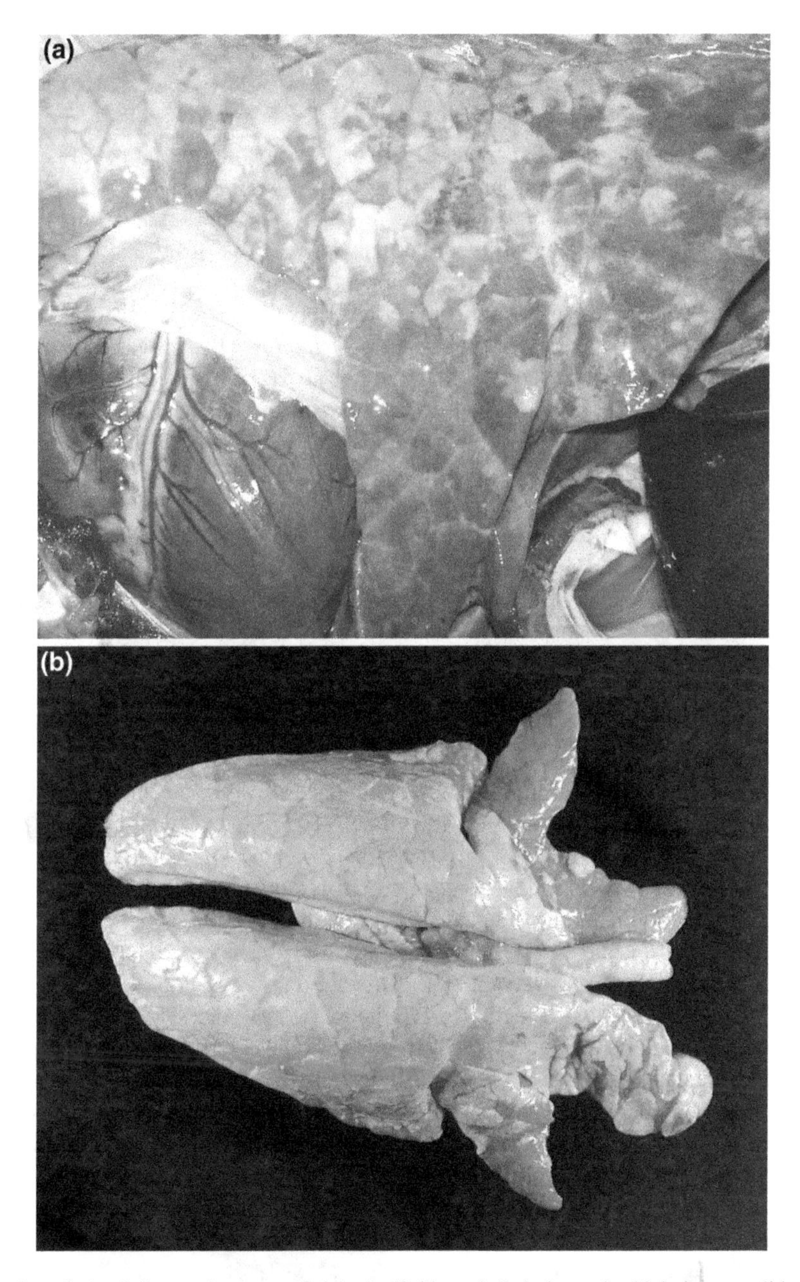

Fig. 1 **a** Swine influenza in a grow-finish pig (field case). Lobular and sublobular consolidation affecting a large portion of cranioventral lung. **b** Lungs from a 6-week-old pig experimentally inoculated with H3N2 SIV and euthanized 5 days postinoculation. Multifocal to coalescing consolidation in cranioventral portions of lung

4.2 Microscopic Lesions

The two most detailed microscopic descriptions of the effects of SIV infection on swine respiratory tract are a histopathologic and immunofluorescent study (Nayak et al. 1965) and an electron microscopic (ultrastructural) study (Winkler and Cheville 1986), both of which are experimental challenge studies with classic H1N1 virus. Less comprehensive but similar descriptions of microscopic lesions induced by SIV infection, some with concurrent immunohistochemical (IHC) studies describing virus distribution in tissues, have been reported by numerous researchers in studies characterizing isolates of interest (Brown et al. 1993; Van Reeth et al. 1996; Landolt et al. 2003; Jung et al. 2005; De Vleeschauwer et al. 2009; Sreta et al. 2009). Additional descriptions of the effect of SIV infection in pigs can also be found in many other studies, often in comparison to human and avian viruses inoculated into swine or in vaccine trials. The descriptions below are composites drawn from these studies as well as the author's experience with both field cases (Janke 1998) and experimental trials (Thacker et al. 2001; Richt et al. 2003, 2006; Solorzano et al. 2005; Vincent et al. 2006, 2007, 2008, 2009a, b; Kitikoon et al. 2009; Ma et al. 2010). The hallmark microscopic lesion of influenza infection, consistently present, is necrotizing bronchitis and bronchiolitis (Fig. 2a). Interstitial pneumonia, though usually evident to some degree, is quite variable in severity in both field cases and in experimental trials, often with pig-to-pig variation.

The earliest response to infection is neutrophil infiltration. By 4–8 h PI, neutrophils are emigrating through airway epithelial layers and accumulating in the lumens of alveolar capillaries. Endothelial cells lining the capillaries are swollen and pavementing of vessel walls by marginated neutrophils is evident. Alveolar walls are widened by vascular congestion and lymphatic dilation. Although ballooning degeneration and cytoplasmic vacuolization of some epithelial cells lining smaller bronchioles may be recognized as early as 8–16 h, these changes are subtle and scattered, and most airways are still intact.

By 24 h PI, extensive infection of epithelial cells lining scattered airways of variable size can be detected by IHC. In a few studies, airway epithelial necrosis was described at this time, but in most studies, disruption of the epithelial layer in a significant number of airways has not yet occurred. Small numbers of neutrophils may be clustered in some airway lumens, accompanied by light infiltration of lymphocytes around some bronchioles. Alveolar walls may be more prominently thickened by congestion, edema, and leukocyte infiltration, predominantly peribronchiolar in distribution.

By 48 h, there is extensive necrosis and sloughing of epithelial cells into airway lumens accompanied by more obvious neutrophil accumulation. Loose infiltration of lymphocytes around affected airways is more prominent but still light. The epithelial cells remaining attached are swollen or attenuated and the layer is irregular in outline. Thickening of alveolar walls, if prominent, is more diffuse. Pneumocytes lining alveoli may be swollen with some sloughing into the lumen.

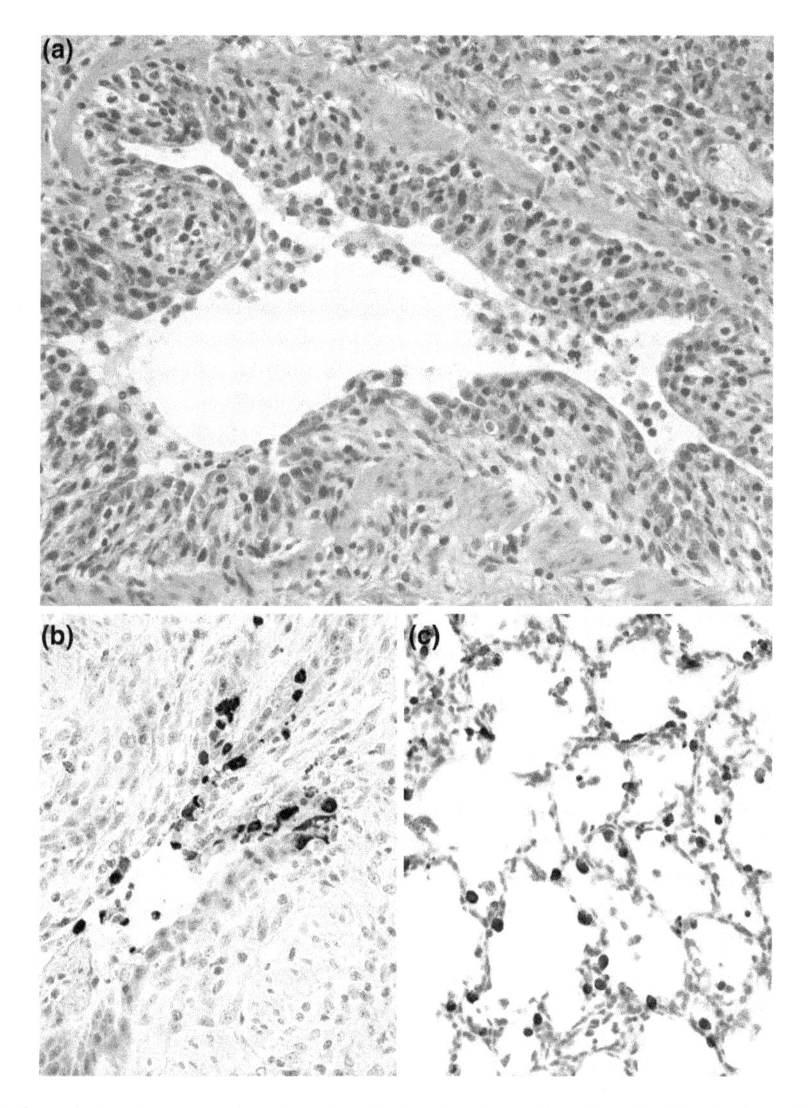

Fig. 2 **a** Subacute necrotizing bronchiolitis in the lung of a 6-week-old pig inoculated intratracheally with H3N2 SIV and euthanized at 3 days postinoculation. Extensive necrosis and sloughing of epithelial cells from a segmental bronchiole is evident. HE x40. **b** Immunohistochemical staining of a similar bronchiole from the same pig identifying virus-infected epithelial cells sloughing into the lumen. IHC x40. **c** Immunohistochemical staining of alveoli from the lung of a 6-week-old pig inoculated with H1N1 SIV by nebulization and euthanized at 24 h postinoculation. Virus has penetrated deep into the lung and numerous pneumocytes lining alveolar walls are infected. IHC x40

Numerous epithelial cells in affected airways (Fig. 2b), both attached and sloughed, contain virus antigen by IHC, but only a few individual to small clusters of infected cells will be observed in alveoli. Some of these cells are obviously swollen pneumocytes still attached to alveolar walls or sloughed into the lumen (Fig. 2c), but other cells within the alveolar wall or loose in the alveolar lumen appear to be macrophages. By close examination, necrotic constituent cells may be identified in alveolar walls, but the septa remain intact. In occasional severe infections, clumps of necrotic cell debris are evident in clusters of alveoli. Consistent with the macroscopic appearance, lobules within the same section of lung may differ in the degree of involvement. Severely affected lobules are frequently atelectatic.

In both field cases and experimental challenge studies, the sizes of airways that are affected may vary. Most likely due to the dynamics of airflow that affect droplet suspension and deposition or to receptor distribution, the largest airways without cartilage (primary bronchioles) are most consistently and most severely affected (Thacker et al. 2001). In some animals, both experimentally and in field cases, the larger lobar or segmental bronchi may be spared. Conversely, in some mild infections, only these larger airways may be infected. The smallest terminal or respiratory bronchioles may be spared or may be necrotic, and the degree of alveolar involvement varies. Except in severe cases, the lesions are multifocal and unaffected lobules sit adjacent to severely affected lobules.

By 72 h PI, some airways are in active necrosis and filled with debris, but many airways are lined by an intact hyperplastic epithelial layer. Peribronchiolar lymphocytic cuffs are well-developed. Alveolar walls may still be thickened as described above with a light loose mixed population of sloughed pneumocytes, macrophages and leukocytes residing in alveolar lumens. Leukocyte populations have shifted to predominantly mononuclear cells. By this time, few infected airway epithelial cells will be identified by IHC but more numerous scattered infected cells, often limited to certain lobules, will be detected in alveoli. In some infections, alveoli may be little affected.

By 96 h and beyond, airways are in repair, lined by hyperplastic or nearly normal epithelium and surrounded by moderate-sized lymphocytic cuffs. Alveolar inflammation is also resolving. By this time, very little virus can usually be detected by IHC, in occasional isolated airways or in scattered individual cells in alveoli. In field cases, in some severely damaged bronchioles, repair is accompanied by fibrosis and endobronchial polyp formation (bronchiolitis obliterans). Such lesions are rarely observed in experimental infections. Over the following days, the epithelial hyperplasia resolves and peribronchiolar lymphocytic cuffing and partially atelectatic alveoli with variable leukocyte populations are all that remain. Though somewhat dependent on the extent of damage, lungs return to normal by 2 weeks PI.

In trachea, infected epithelial cells, as identified by IHC, are usually few in number and widely scattered. Damage to the tracheal epithelial lining, as characterized by attenuation or squamous metaplasia, if present at all, tends to be focal to multifocal. Very few viruses induce extensive epithelial injury, and even then, not consistently in all pigs. Subepithelial lymphocyte infiltration may be intense in the latter situations, but in most tracheas with minimal or focal epithelial attenuation, inflammation exhibits the same range of variation in severity as that observed in pigs not challenged with SIV. Although pigs shed quantities of virus over multiple days in nasal secretions, only mild attenuation of the epithelium lining the inside of nasal turbinate scrolls is observed and that inconsistently. Only low numbers of infected cells are usually identifiable in histopathologic sections by IHC. Infection of tonsil and tracheobronchial lymph node has been reported in some studies described above. In the author's experience, low numbers of infected cells can be identified by IHC in the lymph nodes but rarely in tonsil.

This rapid sequence of events will occur in both the individually affected lobules in most naturally infected pigs and in pigs experimentally infected with high doses of virus. Under experimental challenge conditions in which less virus may be given, examination of multiple sections of lung may reveal asynchronous infection with different lobules at different stages of infection.

5 The Question of Virulence

Although some outbreaks of epidemic swine influenza suggest that certain viruses may be more pathogenic or virulent than others, defining or discovering the basis for virulence has proven to be difficult. Even viruses from severe field outbreaks have tended to be rather tame in captivity (Ma et al. 2010). Most comparative studies conducted in tandem under the same protocols have revealed only minor differences between viruses. More virulent viruses have been deemed so, not so much on increased severity of clinical disease, but on higher or more prolonged fever, higher or prolonged virus shedding or titers in lung, and more extensive macroscopic lesions. Microscopically, lesions tend to be similar but with more lobules and more airways within lobules affected.

Comparative evaluation of results of studies from different researchers must be done cautiously because of the multitude of factors (virus titer, route, and method of inoculation, age of pigs, etc.) that can affect results. Thus, the most significant observations in this regard are likely to come from researchers who have used similar protocols in multiple animal trials with many different viruses. Clinical differences are usually subtle but analysis of parameters that can be quantified can yield clues to virulence differences, e.g. fever, cytokine levels, virus shedding, virus titers in lung tissue or BALF, gross and microscopic lesion scoring. (See Landolt et al. 2003; Richt et al. 2003 for examples of scoring). Although differences may be minimal or below statistical significance because of the number of animals that researchers can reasonably afford to use in such trials, the various parameters often correlate well. In the field situation, such differences would likely be amplified. Some studies suggest the newer triple reassortant viruses may be more virulent than preceding classic viruses (Vincent et al. 2006; Ma et al. 2010). Several recent triple reassortant viruses recovered from situations in which both pigs and people were infected also appear to be slightly more virulent in swine from which they originated (Vincent et al. 2009b).

Most of the research into virulence factors in influenza viruses has been initiated with two unquestionably virulent viruses: H5N1 highly pathogenic avian influenza virus and reconstructed 1918 human pandemic virus (Tumpey et al. 2005; de Wit and Fouchier 2008; de Wit et al. 2008; Basler and Aguilar 2008; Lycett et al. 2009; Janke et al. 2010). The reverse genetics techniques now available allow researchers to replace specific genes or gene sequences in influenza viruses to compare the relative contribution of each gene to virulence or infectivity. Pigs are one of the models (mice, ferrets, chickens, primates) used for such studies, and information gained from such studies may eventually benefit understanding and control of SIV infection (Memoli et al. 2009; Solorzano et al. 2005; Richt et al. 2006).

Acknowledgments The author would like to gratefully thank Dr. Steve Henry, Abilene Animal Hospital, Abilene, KS for his contribution on the current presentation of swine influenza in the field and Dr. Kent Schwartz, Iowa State University, Ames, IA for his contribution of knowledge and insight into all things pig.

References

Basler CF, Aguilar PV (2008) Progress in identifying virulence determinants of the 1918 H1N1 and the southeast Asian H5N1 influenza A viruses. Antiviral Res 79:166–178

Brown IH, Done SH, Spencer YI et al (1993) Pathogenicity of a swine influenza H1N1 virus antigenically distinguishable from classical and European strains. Vet Rec 132:598–602

De Vleeschauwer A, Atanasova K, Van Borm S et al (2009) Comparative pathogenesis of an avian H5N2 and a swine H1N1 influenza virus in pigs. PLoSONE 4(8):e6662. doi:10.1371/journal.pone.0006662

de Wit E, Fouchier RAM (2008) Emerging influenza. J Clin Virol 41:1–6

de Wit E, Kawaoka Y, de Jong MD et al (2008) Pathogenicity of highly pathogenic avian influenza virus in mammals. Vaccine 26 (Suppl 4): D54–D58

Dorset M, McBryde CN, Niles WB (1922) Remarks on "hog flu". J Am Vet Med Assoc 62:162–171

Halbur PG, Paul PS, Frey ML et al (1995) Comparisons of the pathogenicity of two US porcine reproductive and respiratory syndrome virus isolates with that of the Lelystad virus. Vet Pathol 32:648–660

Janke BH (1998) Classic swine influenza. Large Anim Pract 19:24–29

Janke BH, Yoon K-J, Frank R et al (2000) Swine influenza and the porcine respiratory complex. In Keeping pace with SIV: New strategies in diagnostics and management. Proceedings, pre-conference seminar, Allen D Leman Swine Conference, University of Minnesota, St. Paul, MN

Janke BH, Webby R, Lager KM et al (2010) Permissiveness of swine to infection with H5N1 highly pathogenic avian influenza viruses. Proceedings, 52nd Ann Mtg Amer Assoc Vet Lab Diagn, San Diego, CA

Jung K, Ha Y, Chae C (2005) Pathogenesis of swine influenza virus subtype H1N2 infection in pigs. J Comp Pathol 132:179–184

Kitikoon P, Vincent AL, Janke BH et al (2009) Swine influenza matrix 2 (M2) protein contributes to protection against infection with different H1 swine influenza virus (SIV) isolates. Vaccine 28 (2):523–531.

Landolt GA, Karasin AI, Phillips L et al (2003) Comparison of the pathogenesis of two genetically different H3N2 influenza A viruses in pigs. J Clin Microbiol 41:1936–1941

Lycett SJ, Ward MJ, Lewis FI et al (2009) Detection of mammalian virulence determinants in highly pathogenic avian influenza H5N1 viruses: Multivariate analysis of published data. J Virol 83:9901–9910

Ma W, Vincent AL, Lager KM et al (2010) Identification and characterization of a highly virulent triple reassortant H1N1 swine influenza virus in the United States. Virus Genes 40:28–36

McBryde CN (1927) Some observations on "hog flu" and its seasonal prevalence in Iowa. J AmVet Med Assoc 71:368–377

Memoli MJ, Tumpey TM, Jagger et al (2009) An early 'classical' swine H1N1 influenza virus shows similar pathogenicity to the 1918 pandemic virus in ferrets and mice. Virol 393:338–345

Nayak DP, Twiehaus MJ, Kelley GW et al (1965) Immunocytologic and histopathologic development of experimental swine influenza infection in pigs. Am J Vet Res 26:1271–1283

Richt JA, Lager KM, Janke BH et al (2003) Pathogenic and antigenic properties of phylogenetically distinct reassortant H3N2 swine influenza viruses co-circulating in the United States. J Clin Microbiol 41(7):3198–3205

Richt JA, Lekcharoensuk P, Lager KM et al (2006) Vaccination of pigs against swine influenza virus by using an NS1-truncated modified live-virus vaccine. J Virol 80 (22):11009–11018

Solorzano A, Garcia-Sastre A, Richt JA et al (2005) Mutations in the NS1 protein of swine influenza virus impairs anti-interferon activity and confers attenuation. J Virol 79:7535–7543

Sreta D, Kedkovid R, Tuamsang S et al (2009) Pathogenesis of swine influenza virus (Thai isolates) in weanling pigs: An experimental trial. Virol J 6:34 doi:10.1186/1743-422X-6-34

Thacker EL, Thacker BJ, Janke BH (2001) Interaction between Mycoplasma hyopneumoniae and swine influenza virus. J Clin Microbiol 39:2525–2530

Tumpey TM, Basler CF, Aguilar PV et al (2005) Characterization of the reconstructed 1918 Spanish influenza pandemic virus. Science 310:77–80

Van Reeth K, Pensaert MB (1994) Porcine respiratory conronavirus-mediated interference against influenza virus replication in the respiratory tract of feeder pigs. Am J Vet Res 55:1275–1281

Van Reeth K, Nauwynck H, Pensaert MB (1996) Dual infections of feeder pigs with porcine reproductive and respiratory virus followed by porcine respiratory coronavirus or swine influenza virus: A clinical and virological study. Met Microbiol 48:325–335

Van Reeth K, Labarque G, De Clercq S et al (2001) Efficacy of vaccination of pigs with different H1N1 swine influenza viruses using a recent challenge strain and different parameter of protection. Vaccine 19:4479–4486

Vincent AL, Lager KM, Ma W et al (2006) Evaluation of hemagglutinin subtype 1 swine influenza viruses from the United States. Vet Microbiol 118:212–222

Vincent AL, Ma W, Lager KM et al (2007) Efficacy of intranasal administration of a truncated NS1 modified live influenza virus vaccine in swine. Vaccine 25:7999–8009

Vincent AL, Lager KM, Janke BH et al (2008) Failure of protection and enhanced pneumonia with a US H1N2 swine influenza virus in pigs vaccinated with an inactivated classical swine H1N1 vaccine. Vet Microbiol 126:310–323

Vincent AL, Ma W, Lager KM et al (2009a) Characterization of a newly emerged genetic cluster of H1N1 and H1N2 swine influenza virus in the United States. Virus Genes. doi 10.1007/s11262-009-0386-6

Vincent AL, Swenson SL, Lager KM et al (2009b) Characterization of an influenza A virus isolated from pigs during an outbreak of respiratory disease in swine and people during a county fair in the United States. Vet Microbiol 137:51–59

Winkler GC, Cheville NF (1986) Ultrastructural morphometric investigation of early lesions in the pulmonary alveolar region of pigs during experimental swine influenza infection. Am J Pathol 122:541–552.

Diagnostics and Surveillance for Swine Influenza

Susan Detmer, Marie Gramer, Sagar Goyal, Montserrat Torremorell and Jerry Torrison

Abstract Collective knowledge regarding the occurrence of influenza among swine is incomplete due to inconsistent surveillance of swine populations. In this chapter, we review what surveillance activities exist and some of the practical challenges encountered. Furthermore, to support robust surveillance activities, accurate laboratory assays are needed for the detection of the virus and viral nucleic acids within clinical samples, or for antiviral antibodies in serum samples. The most common influenza diagnostic assays used for swine are explained and their use as surveillance tools evaluated.

Contents

S. Detmer · M. Gramer (✉) · S. Goyal · M. Torremorell · J. Torrison
Department of Veterinary Population Medicine, College of Veterinary Medicine,
University of Minnesota, Saint Paul, MN, USA
e-mail: grame003@umn.edu

Current Topics in Microbiology and Immunology (2013) 370: 85–112

DOI: 10.1007/82_2012_220

Published Online: 8 May 2012

1 Surveillance for Influenza Viruses in Swine

Influenza A viruses in swine typically cause an acute respiratory disease which, in uncomplicated cases, is mild and self-limiting (Radostits et al. 2000). Infection of swine with influenza A virus is common (Brown 2000) and occurs throughout the year (Vincent et al. 2008). However, seasonal peaks occur in months with moderate temperatures and humidity (Shaman and Kohn 2009) similar to the pattern of disease seen in humans. Because endemic swine influenza is highly prevalent but causes minimal mortality in infected pigs, the World Organization for Animal Health (Office International de Epizooties), and the U.S. Department of Agriculture (USDA) have not classified swine influenza as a notifiable or reportable disease (OIE 2009; USDA/APHIS 2009). Further complicating the coordinated surveillance efforts are the limited resources available for animal disease surveillance in general. For these financial and biological reasons, systematic and rigorous surveillance is focused on diseases of much higher consequence to animal health and international trade, such as brucellosis and foot-and-mouth disease. Animal disease surveillance in general is labor-intensive and costly and hence animal health authorities at the international, national, provincial, and state levels have precluded assigning it a higher priority for funding (Pappaioanou and Gramer 2010). Given these challenges, the efficient and effective surveillance of influenza viruses in swine will require a strategic approach, encompassing all the attributes of a successful surveillance program.

1.1 Attributes of Disease Surveillance Systems

When considering an influenza virus surveillance program for swine populations the key attributes of disease surveillance systems developed and used by leading public health authorities for detecting diseases of public health importance in human populations (CDC 2009) must be considered. These attributes are summarized below.

A. *Simplicity*. This refers to the surveillance system's structure and ease of operation. As with most successful operations, systems that prove to be the most valuable utilize methods that are as simple as possible while still fulfilling the primary objectives.
B. *Flexibility*. A system that can adapt to changing needs, such as the addition of new collection methods or employing new and more specific diagnostic assays has built-in flexibility to capture the required information.
C. *Acceptability*. A surveillance system must appeal to all interested parties and, once found acceptable, it reflects the willingness of individuals and organizations to participate in the surveillance system (e.g., swine farmers, veterinarians, and veterinary diagnostic laboratory personnel who are asked to report cases of disease).

D. *Timeliness.* After initial diagnosis, how quickly the cases are entered into the surveillance system or the time that elapses between onset of infection, diagnosis, case report, information sharing, and action, is often regarded as key to a surveillance system's success (Jajosky and Groseclose 2004). While timeliness is of critical importance, it is often very difficult to measure (Jajosky and Groseclose 2004).
E. *Completeness.* Completeness is the attribute of a surveillance system that is most directly linked to the true discipline of epidemiology. Completeness is reflected by the proportion of all cases of disease in a specified population that are detected by the surveillance system and is affected by the likelihood that: (a) animals with infection or disease are tested; (b) the condition is correctly diagnosed (skill of animal health provider, accuracy of diagnostic tests); and (c) the case is reported to the surveillance system once it has been diagnosed. The factors that may affect completeness of a surveillance system are addressed in more detail in Sect. 1.3.
F. *Representativeness.* A surveillance system that accurately describes the occurrence of disease over time and its distribution in the study population by location, group, and severity can be referred to as representative. Consideration of this attribute is especially important for large populations with variable prevalence because most systems simply cannot detect every single case of infection or disease. The common idiom "tip of the iceberg" is a popular way of referring to how the documented and described cases of a disease that are evident as the result of a surveillance program truly represent the largely undetected/unseen cases in the vast population.

While there are several challenges that inherently exist when trying to conduct surveillance on animal populations (further discussed in Sect. 1.3), if an organization is forward thinking and keeps these key attributes in mind, a wealth of information can be generated. The information garnered from an influenza surveillance system for pigs is driven in large part by the initial design and rationale for the surveillance.

1.2 Rationale Behind Influenza Surveillance Systems for Swine

While the majority of a surveillance effort is designed to answer "how" and "what", e.g., logistics and population selection, the better part of the planning time should be devoted to determining "why" to invest in the activity in the first place. For influenza virus surveillance in animals, the simple, altruistic reason is that surveillance must be done to protect public health and prevent pandemics (Patriarca and Cox 1997). Philanthropic intentions aside, it is important to find rationale for surveillance that will also benefit or provide information to all stakeholders. A recent Institute of Medicine (IOM) review for the National Academies, "Sustaining Global Surveillance and Response to Emerging Zoonotic

Diseases," hits the nail on the head with its recommendations to improve early detection and response to zoonotic diseases, such as influenza. Specifically, they are of the opinion that comprehensive surveillance would be best achieved in the following manner: "Multidisciplinary teams of professionals that have relevant expertise and field experience would identify populations at risk and causes and risk factors for infection, and then rapidly and widely disseminate this information so that immediate and longer term disease prevention and control interventions can be implemented (IOM/NRC 2009)."

For influenza virus surveillance in swine in the United States, the rationale for a surveillance system includes not only protection of public health, but detection, discovery, and sharing of virus isolates to facilitate updates for vaccines, refine diagnostic assays, and determine the distribution of new influenza strains in swine to inform further policy decisions (USDA/APHIS 2009). In Europe, the Research Programme of the European Commission funded the coordination of the European Surveillance Network for Influenza in Pigs (ESNIP), a group that set out on a coordinated surveillance mission many years earlier, in 2001, with the stated goal of being to first standardize diagnostic techniques used for surveillance and detection of influenza viruses in pigs. Once the initial goals were achieved, the wealth of information from the surveillance efforts was leveraged for a second round of studies (ESNIP 2007) on the epidemiology and evolution of influenza viruses in European pigs and to optimize influenza diagnostic assays for swine (Kyriakis et al. 2010a). Naturally, a listed rationale for ESNIP 2 is, "to obtain insights into the public health risk of influenza in swine by monitoring swine for avian influenza viruses and by comparison of influenza viruses in swine and in human populations." ESNIP 3 has since been launched to "…increase the knowledge of the epidemiology and evolution of swine influenza virus in European pigs" with significant research investment directed toward detailed antigenic and genetic characterization of influenza virus strains isolated from pigs (European Commission 2010).

The surveillance programs for swine influenza in developed countries such as the United States and those of the European Union are striving to best coordinate efforts not only with multiple disciplines and agencies, but also with swine producers. While the benefits to human and public health are tangible in that understanding influenza virus patterns in swine may lead to more accurate and timely diagnosis of zoonotic influenza events, the reward for swine producers is less definable. The perception that influenza surveillance programs for pigs have fewer advantages for pork producers is likely due to several reasons, including the minimal impact of influenza virus infection on overall swine health and productivity, fear that trade and profits will be negatively affected, and the lack of readily available, consistently reliable, and inexpensive vaccines to control influenza once it is detected. Furthermore, the funding for such surveillance efforts in pigs requires commitment from all sectors, including animal agriculture, food production, human, and public health (AASV 2009). Securing funding and increasing participation in influenza surveillance programs for swine are challenges that need to be addressed.

1.3 Challenges to Surveillance in Animal Populations

From the outset, any effort to conduct influenza surveillance in pigs faces several unique challenges. For starters, respiratory disease is relatively common in pigs and the clinical signs and gross lesions associated with influenza virus in pigs are not entirely specific to influenza. While influenza virus is a significant disease in pigs, there is no official disease reporting requirement for influenza because clinical disease seldom leads to dramatic mortality or severe economic losses in a herd. Additionally, influenza virus is a highly mutagenic virus that can be exchanged among multiple species, with most concerning exchanges occurring between animals and humans. Due to the nature of the global economy, both humans and animals are increasingly mobile regionally and internationally, making comprehensive surveillance difficult across species and geographic boundaries. These factors pose challenges to any influenza surveillance effort in pigs and illustrate the importance of a coordinated surveillance approach.

The first challenge to swine influenza surveillance, and to any early warning system for swine infectious disease, is the inability to reliably detect disease through observations of clinical signs. The clinical signs associated with influenza virus in pigs are generally attributed to Porcine Respiratory Disease Complex (PRDC), a polymicrobial pneumonia caused by several common swine respiratory viruses and bacteria (Brockmeier et al. 2002; Straw et al. 2006). While it is true that influenza A viruses are frequently isolated from pigs with PRDC and evidence of exposure via serological assays is common in growing swine (Brockmeier et al. 2004), it would be wrong to ascribe all clinical signs of respiratory disease in a swine population to influenza without more discriminatory diagnostic methods. Therefore, while reliance on clinical signs and gross lesions for disease detection in pigs has proven to improve the sensitivity of disease detection for other swine diseases such as classical swine fever (Elbers et al. 2003), the sole use of clinical signs as a detection method can lack specificity (Engel et al. 2005). Even swine diseases with hallmark clinical signs, e.g. vesicular exanthema of swine, have cases that may be mild enough to fail detection by clinical observation alone (Schnurrenberger et al. 1987). Similarly, many uncomplicated influenza virus infections in pigs are also mild. It has been shown that single virus infections result in transient clinical signs (Van Reeth et al. 1996). Hence, clinical signs as a method of detection for influenza virus would also likely result in numerous missed cases as well as an abundance of false positive cases. Finally, using clinical signs to detect influenza requires observation by a trained veterinary professional or animal caretaker. Thus, it bears mentioning that surveillance methods, like direct clinical observation, requiring close contact with animals infected by with a potentially zoonotic disease can pose health risks to workers, another potential challenge (Myers et al. 2006; Bos et al. 2010).

The challenges associated with tracking "transboundary" viruses in animals, including influenza, have been reviewed previously (Domenech et al. 2006; Lynn et al. 2006; Gubernot et al. 2008) and the impact of human travel on respiratory disease epidemics such as Severe Acute Respiratory Syndrome (SARS) in people has been examined extensively. However, the component of virus transmission from humans to pigs has not been a significant consideration other than in retrospective analysis of the 1918 Spanish Flu pandemic up until 2009 H1N1 pandemic (Hofshagen et al. 2009). Clearly the impact of human travel and the potential for infecting pigs with novel influenza viruses are evident now. Yet few surveillance systems that exist are capable of capturing both the human and animal data needed to shed light on the existing barriers that prevent or gateways that allow transmission to occur between species.

Migratory waterfowl represent another potential transmission source for influenza to pigs, as demonstrated experimentally (Kida et al. 1994) and naturally (Pensaert et al. 1981; Karasin et al. 2000). In the more recent report on natural infection of pigs with H4N6 influenza, waterfowl on a lake near a swine farm in Canada were implicated as the source of infection in pigs (Karasin et al. 2000). Even with confinement rearing of pigs, exposure to water-borne virus is possible in cases where surface water is used untreated as a water supply for the pigs. Pigs raised partially or completely outdoors could face a higher exposure risk. In the case of the H4N6 influenza virus infection, pigs were raised in confinement. The authors provide evidence of pig to pig transmission of the H4N6 influenza virus within the herd.

In contrast, the first widespread detection of H3N2 influenza virus in pigs in the United States in 1998 was followed by widespread dissemination of H3N2 throughout the North American swine population and subsequent reassortment with other influenza viruses (Ma et al. 2006). This is significant in light of the tremendous increase in movement of growing pigs throughout North America over the past 20 years, another significant challenge for surveillance of influenza in pigs. Data from the Minnesota Board of Animal Health on the movement of growing pigs and breeding swine into Minnesota are illustrative of this point (Fig. 1). There has been more than a seven fold increase in the number of feeder pigs imported into Minnesota over a 5-year period (fiscal years 1994–1999) and this number has doubled again in the subsequent 5-year period (fiscal years 1999–2004), with shipments originating from Canada and 31 other U.S. states (source: Minnesota Board of Animal Health). This movement of pigs at young ages (3–11 weeks) provides a source of pigs that are potentially infectious or susceptible (or both) to particular influenza virus strains. This extent of interstate and international movement is an important consideration when designing surveillance methods for influenza in pigs.

Finally, there is a potential challenge to influenza surveillance in pigs if producers are reluctant to participate in such a program. Diagnostic testing costs can be a barrier to surveillance particularly during protracted periods of unprofitable production such as occurred in 2009 in North America. Producers and veterinarians may also be reluctant to participate in surveillance programs that are perceived to have a potential negative impact on marketability of pigs from a specific site or more generally for the marketing of pork.

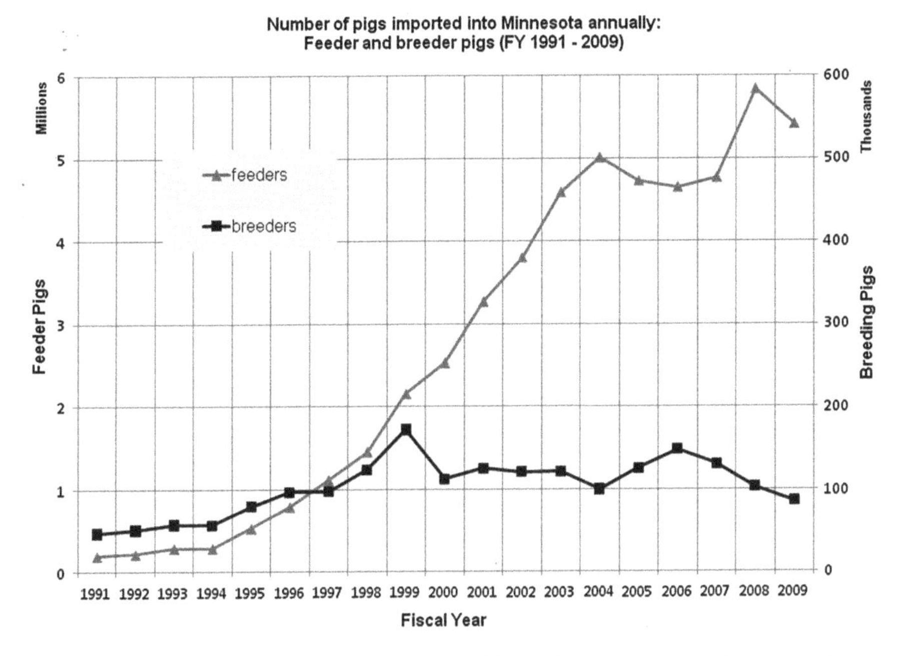

Fig. 1 Number of growing and adult pigs imported into Minnesota annually during fiscal years (FY) 1991 and 2009 according to the Minnesota Board of Animal Health. *Triangles* represent growing (feeder) pigs. *Squares* represent adult (breeder) pigs

1.4 Surveillance Design and Logistics

Surveillance design parameters depend on the objectives of the surveillance program as outlined previously. For example, surveillance parameters would be different if the objective is to identify the most prevalent influenza virus subtypes in pigs in a particular region versus whether influenza virus has been eliminated from a specific swine herd (Torremorell et al. 2009). Designing a surveillance program also requires a thorough understanding of the behavior of the virus in pigs, available diagnostic tests, and the production practices used for raising pigs that are to be monitored. Important features of influenza virus infections in pigs are illustrated in Fig. 2 and discussed in detail in other chapters (Clinicopathological Features of Swine Influenza) in this text.

Briefly, it is critical to remember that pigs develop a fever and begin shedding virus rapidly following exposure to influenza virus. Peak virus excretion follows the peak of fever very closely and declines rapidly thereafter. Circulating antibodies are detected within 10–14 days of infection. On an individual pig basis, there is a window of time following infection in which the virus has been cleared, antibodies have not developed, and the pig appears not infected.

Surveillance design is also a function of the tests available for use. Tests intended to detect virus need to be applied during the first week following infection,

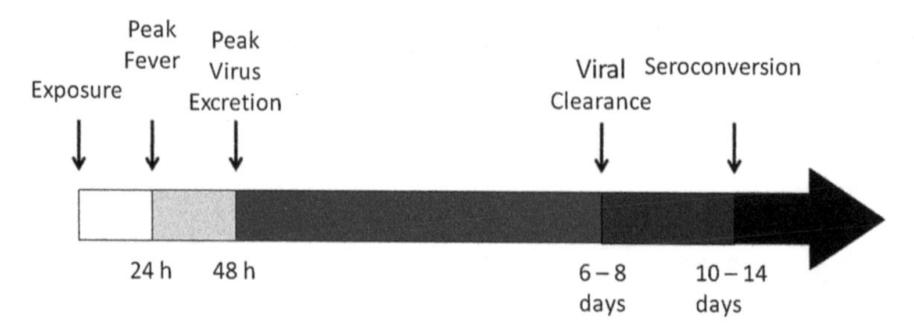

Fig. 2 The dynamics of influenza virus infection in swine represented by the simple timeline here are useful for designing surveillance testing protocols

preferably on samples from pigs that are still febrile. Serological tests such as hemagglutination inhibition can be used to evaluate samples before and after the expected time of seroconversion to specific subtypes of influenza. Serological tests are also available in an ELISA format that detect antibodies against all influenza A subtypes (Ciacci-Zanella et al. 2010) or certain individual subtypes. Influenza diagnostic assays for both antigen and antibody detection are discussed in detail in Sect. 2. Once established diagnostic assays are chosen for the surveillance program, the next critical component is a proper specimen selection and sampling strategy.

Specimen selection and sampling strategies. The specimen of choice within a surveillance program again relates to the objectives of the surveillance as well as the availability of appropriate samples for collection and testing. A variety of specimens are suitable for SIV detection in pigs, including nasal swabs, tracheal swabs, tracheal fluid, lung lavage fluid, and lung sections. For ante-mortem diagnosis of SIV, nasal swabs are one of the more easily obtainable samples. Oral fluids collected from pigs on a group basis represent an alternative to nasal or oropharyngeal swabs. Oral fluids have been used extensively for diagnostic tests in human medicine and are now being applied in swine herds for detecting pathogens and antibodies against the pathogens (Prickett and Zimmerman 2010). Specific applications of oral fluids for influenza virus testing are discussed in Sect. 2.2. Additionally, testing air samples for the presence of swine influenza virus is in the early stages of development (Hermann et al. 2006) and could find application in broader surveillance applications. Postmortem examinations of pigs infected with influenza A viruses have detected the virus (Vincent et al. 2009a; Yazawa et al. 2004) primarily in respiratory tract tissues (nasal turbinates, trachea, and lung), but also in tonsil and bronchial lymph node. The sites for virus replication are similar for historical isolates of "classical" H1N1 swine influenza virus (Yazawa et al. 2004) and 2009 pandemic H1N1 (Vincent et al. 2009a). Postmortem tissues are considered ideal specimens as they can also be examined for gross and microscopic lesions. Thus, complete necropsies with histopathological examinations can further our understanding of the pathogenesis of influenza A viruses in swine. Regardless of the specimen collected, the sample size chosen from the population of concern will affect successful detection of influenza in swine.

Sample size determination for surveillance programs is a function of what test is to be employed and how *prevalent* the target organism is within the population. In other words, the number of pigs shedding influenza virus at the time of sampling is likely to be different from the number of pigs with serum antibodies depending on when the pigs are sampled. The calculation of an adequate sample size required is fairly straightforward once all the other elements of the sampling frame are established, i.e. sensitivity and specificity of the test, prevalence of the target organism within the population, population size to be sampled, and desired confidence in the end result.

A formula that has been used extensively in swine disease surveillance programs for many years is given below:

$$n = (1 - (1 - \alpha)1/d)(N - (d/2)) + 1$$

where N is the population size, d is the number of positives in the population (expected or threshold prevalence for detection), α is the desired confidence level, and n is the number needed for testing (Cannon and Roe 1982). For example, if one assumes that a diagnostic assay is 100 % sensitive and specific, a sample size of $n = 30$ from a herd of infinite population N, will provide a 95 % confidence level of detection of disease if the disease prevalence is 10 %. In most situations, the diagnostic assays employed are not 100 % sensitive. As sensitivity decreases, the sample size n must increase.

Databases and sharing of information. Our experience with swine influenza databases indicates that populating the database and sharing the information is most successful when the information is used for a specific and important purpose. Testing for swine influenza virus is a regular activity at veterinary diagnostic laboratories. Serology results are generally used for making decisions on vaccine timing and are not typically collated into common databases. Virus detection by PCR and virus isolation is used to determine the role of influenza virus in clinical disease. Viruses isolated from clinical cases are often used for the production of autogenous vaccines based on the results of virus sequencing information. Virus sequencing information is often assembled into dendrograms to follow virus trends over time and geography. Each piece of the collective diagnostic information has a role in influenza diagnosis and control on a herd or production system basis. By definition, this brings the maintenance of the database close to the end user, who also happen to provide the data inputs.

Financial incentives, such as third party payment for sequencing information, have not appeared to be as important as the direct need for information in terms of motivating producers to participate in surveillance programs up to now. The degree to which producers and veterinarians are interested in sharing disease surveillance information among groups is promising but has not yet been fully determined (Davies et al. 2007).

1.5 *Examples of Influenza Surveillance in Swine*

Comprehensive surveillance programs are needed to detect new influenza strains especially the ones with pandemic potential so we can increase our preparedness to it. Effective surveillance programs should include detection of influenza viruses in humans and animals including pigs. It should also include detection of viruses distributed throughout the world particularly in high risk areas where humans, poultry, and pigs coexist. Surveillance in pigs is considered crucial because pigs have receptors for human, swine, and avian influenza viruses potentially favoring the arising of new viral reassortants. Unfortunately such a global comprehensive surveillance program has not been put in place yet but attempts have been made at the local and regional levels. One limitation of this approach is that the information is not always integrated and shared across species and regions diminishing the effectiveness of comprehensive surveillance efforts.

Detection of influenza viruses in other mammalian species such as cats, dogs, bovine, and equine should also be considered as part of the integrated programs. Although a coordinated global surveillance initiative in pigs does not exist yet, there are examples of programs that over the years have provided a significant but incomplete picture of the circulating influenza viruses in pigs. In addition, the programs that are being planned to actively collect data and specimens for influenza will help to bridge the current gap in influenza surveillance in pigs.

Serosurveillance of pigs in North America. In the US, surveillance studies using serological methods have been based on the sampling of pigs at the point of slaughter and the testing of samples submitted to veterinary diagnostic laboratories. In these studies, pigs originated from various Midwestern States and were representative of pigs owned by multiple enterprises throughout the US. This was the method of choice for many years when other methods of sampling were not available.

In the US, several serological surveys have been conducted. It was demonstrated during 1976/1977 (Hinshaw et al. 1978) and 1988/1989 (Chambers et al. 1991) that influenza virus infections were common among pigs. The percentage of pigs seropositive against classical swine H1N1 viruses ranged from 20 to 47 % in 1976/1977 and 51 % in 1988/1989. In contrast serologic evidence of H3 virus exposure was remarkably lower in both studies (1.4 % in 1976/1977 and 1.1 % in 1988/1989).

In a subsequent study conducted in 1997/1998 (Olsen et al. 2000), 27.7 % of pigs were seropositive to swine H1 virus, 8 % to an H3 human virus, and 7.6 % to an H1 avian virus. These results indicated that pigs were exposed to human H3 and avian viruses to a greater extent than in the past. The finding that the study population tested positive to human H3 influenza virus was of particular significance. Up to that point, detection of H3-subtype influenza viruses in US pigs was rare although it was detected regularly among pigs in Asia and Europe. The findings in 1998 indicated a dramatic pattern change for influenza epidemiology in North America. A Canadian study indicated that seroprevalence to H3N2 viruses in 2002 was negligible although seroprevalence to H1N1 remained high (24.3–61.1 %) (Poliak et al. 2008).

Therefore, influenza surveillance using serological methods has provided useful information in the past but its use has become less reliable due to the broader use of influenza vaccines in pigs and the inability to differentiate antibodies induced by vaccine strains from field strains. In addition, serological methods may not always be able to differentiate infection by strains within a subtype or even between subtypes. The limitations of serological assays are discussed in further detail in Sect. 2.3. For these reasons, virus molecular characterization methods have become widely used and are better able to detect genetic differences among viruses.

Surveillance provided by U.S. Veterinary Diagnostic Laboratories. State and private diagnostic laboratories in the US constitute a rich resource of samples and data for influenza virus surveillance in pigs. Thousands of cases are submitted to the diagnostic laboratories by practitioners and producers to investigate respiratory disease. Most of the cases originate from US herds but may include samples from Canadian herds and a few countries located in Central and Latin America. In many of the cases submitted, influenza virus is detected and diagnosed. As an example, more than 4862 influenza A viruses have been isolated from swine respiratory specimens at the University of Minnesota Veterinary Diagnostic Laboratory (UMVDL) between January 1, 2001 and June 1, 2010 (Gramer and Torrison 2010). In addition 200–700 influenza A virus nucleic acid detection tests (RT-PCR) are conducted monthly on swine respiratory specimens submitted to the UMVDL. The detection of influenza A virus by PCR is followed by subtyping and even partial hemagglutinin gene sequencing when funding is available. Because of confidentiality issues, data derived from these diagnostics is reported solely to the submitting veterinarian and animal owner. While some of the data is shared with the influenza research community, the majority is not automatically released to any publically accessible surveillance databases. Rectifying this situation is not straightforward, but would likely involve discontinuing the institutional practice of considering animal influenza virus isolates as the *intellectual property* of the owner or researcher and assuring anonymity and prevention of penalties to clients submitting specimens. Nevertheless, diagnostic laboratory data do constitute a valuable resource. In the US, the data generated represent the types of influenza viruses circulating in domestic swine and has resulted in vaccine strain updates and diagnostic reagent revision.

It can be argued that surveillance conducted through routine submissions to diagnostic laboratories is passive, syndromic, and retrospective producing only partial analysis of viruses. Whole genome sequencing of influenza A virus isolates from pigs is needed to detect virus changes and reassortment events that may result in new strains of pandemic potential (Vijaykrishna et al. 2010). Efforts, such as that of the USDA, are designed to integrate the US veterinary diagnostic laboratory network influenza detection and characterization into a more integrated and comprehensive surveillance plan (USDA/APHIS 2009).

Passive/Syndromic surveillance programs. During the last few years discussions have taken place in the US to have an active surveillance influenza program in pigs similar to those for people and poultry for detecting high pathogenic avian influenza or detecting strains of clinical importance. Such a program has not yet been fully possible in pigs although tremendous advances have been made. As a result of the 2009 pandemic, the USDA in cooperation with the CDC and industry allies initiated a voluntary influenza surveillance program in pigs (USDA/APHIS 2009). Although participation in the program has been limited, pork producer, and veterinarian involvement is slowly increasing and contributions of specimens for virus isolation to the surveillance efforts are on the rise.

In addition, influenza is proposed to be part of a comprehensive and integrated surveillance program being designed to protect the US food supply from the impact of diseases considered exotic in the US (AASV 2010). This program has many goals including actively testing for foot-and-mouth disease, classical swine fever, *Brucella suis*, Aujeszky's disease, *Trichinella spiralis*, *Toxoplasma gondii,* and influenza A virus. Many stakeholders are participating in the design of this program, including the USDA, HHS/CDC, National Pork Board, National Pork Producers Council, American Association of Swine Veterinarians, Veterinary Diagnostic Laboratories, State Animal Health Officials, and State Pork Producer Associations. In regards to influenza, the program aims to determine the prevalence and variety of influenza viruses in US swine, facilitate influenza strain selection for vaccine production, provide continuous improvement of diagnostic testing capabilities, and warrant anonymity to the submitting systems to facilitate cooperation. Such a system should facilitate the cooperation and sharing of information and specimens among stakeholders.

Hong Kong surveillance program for influenza in slaughtered swine. For over a decade, researchers at the University of Hong Kong have participated in an internationally funded, systematic, virological, surveillance program for influenza A viruses in swine slaughtered at one abbatoir in Hong Kong (Vijaykrishna et al. 2010). A majority of the swine slaughtered at this abbatoir are said to originate from mainland China. Routine visits are made to the abbatoir wherein nasal or tracheal swabs from slaughtered pigs are collected, subjected to virus isolation via inoculation in eggs or MDCK cells, and then characterized by hemagglutination inhibition (HI) and sequencing. This slaughter surveillance program has yielded interesting information regarding the genetic constellation of viruses present in China and Hong Kong (Smith et al. 2009; Peiris et al. 2001).

Research-based surveillance. In an effort to bridge the gap on influenza surveillance in pigs, the United States National Institutes of Health funded Centers of Excellence for Influenza Research and Surveillance (CEIRS) have directed some of their research efforts toward active influenza surveillance in swine-dense areas in the Midwestern United States (NIAID 2010). The information from an active surveillance program such as this is sorely needed as growing swine are more representative of the population of pigs most likely to be infected with influenza A virus (Brown 2000), and, because the epidemiology of the virus in swine farms is not well understood (Olsen et al. 2006), an active surveillance program can shed key information on the epidemiology of influenza in swine. In the NIAID sponsored program on active influenza surveillance in swine, thirty nasal swabs are collected every month for 12 consecutive months from growing pigs in 34 separate farms. Swabs are tested for influenza virus by PCR and virus isolation. During collection, the age of the pigs, group clinical signs, and influenza vaccination history are recorded. Farm characteristics, such as herd size, building design, proximity to other farms, biosecurity practices, are also recorded in an attempt to determine possible risk factors associated influenza virus infection. Data on pig age, clinical status, meteorological, and environmental conditions are collected to obtain information on current influenza isolates, their distribution, and disease characteristics.

Summary of international surveillance programs. In Europe, the Research Programme of the European Commission funded the coordination of the European Surveillance Network for Influenza in Pigs (ESNIP). This group became active in 2001 and continues the efforts to increase the knowledge of the epidemiology and evolution of swine influenza virus in European pigs.

In Hong Kong, the surveillance program consists of the isolation of influenza virus at the point of slaughter. Throughout this program a limited but significant number of viral isolates has become available representing the only active systematic influenza surveillance program in the world.

In South and Central America, formal surveillance efforts are nonexistent and are complicated by the fact that some countries consider influenza in pigs an exotic disease limiting the ability to even conduct routine influenza diagnostics.

2 Diagnostics for Swine Influenza

Diagnosis of swine influenza in the twenty-first century has become more complicated due to the presence of multiple strains of influenza viruses cocirculating in pigs (Webby et al. 2004). Due to the introduction of these multiple strains, the diagnosis and characterization, it is important to understand the many tests that are being used to better characterize influenza virus infections in swine.

2.1 Clinicopathology

Clinical signs and characteristic macroscopic and microscopic lesions are useful in making a presumptive, but not definitive, diagnosis of swine influenza infection (see the chapter regarding Clinicopathological Features of Swine Influenza in this text and also Sect. 1.3 and Fig. 2). Laboratory detection of the whole virus, viral antigen, viral nucleic acids or anti-viral antibodies within tissues, serum or other clinical samples is needed for definitive diagnosis.

2.2 Direct Detection Methods

2.2.1 Detection of Influenza Virus Antigen

Immunohistochemistry (IHC) and immunofluorescence (IFA) are used to detect influenza virus antigen in frozen or formalin-fixed tissues using different antibodies (Guarner et al. 2000; Haines et al. 1993; Larochelle et al. 1994; Onno et al. 1990; Vincent et al. 1997). The nucleoprotein (NP) is well-conserved among influenza A viruses; therefore, anti-NP antibodies can be used to detect all subtypes of influenza A viruses. However, the hemagglutinin (HA) protein is subtype-specific and hence is used to detect specific subtypes of influenza virus. The NP antigen is located in the nucleus and cytoplasm of infected cells (Guarner et al. 2000; Haines et al. 1993; Larochelle et al. 1994; Vincent et al. 1997) while the HA is located in the cytoplasm and along the cell surface (Guarner et al. 2000).

Direct immunostaining methods use antibodies that are labeled with biotin, fluorophore, enzyme, or colloidal gold (Buchwalow et al. 2010). Although technically difficult and time-consuming, indirect immunostaining methods have higher sensitivity and are more commonly used for diagnostic tests (Buchwalow et al. 2010). These methods use an unlabeled primary antibody followed by a labeled secondary antibody. The application of the substrate then results in amplification of the colorimetric signal produced by the enzyme attached to the secondary antibody (Buchwalow et al. 2010). Of the indirect methods, the standard avidin–biotin complex (ABC) method of IHC has been widely used for SIV detection (Haines et al. 1993; Vincent et al. 1997). However, with this method there can be background staining due to endogenous biotin in the tissues (Vosse et al. 2007). Therefore, these methods have been adapted to polymer-based IHC method (Richt et al. 2006) that uses a polymer backbone on the secondary antibody to attach to the enzyme instead of avidin–biotin complex (Sabattini et al. 1998).

A number of rapid immunoassays, most being enzyme-linked immunosorbent assay (ELISA)-based tests kits are commercially available that can detect influenza virus antigen in clinical samples. Most of these tests have been developed specifically for human and avian applications and the viral proteins that are detected by these kits are HA, neuraminidase (NA), or NP. Five of the kits licensed for human application were found to have sensitivity of 67–71 % and specificity of 99–100 % for Influenza A (Hurt et al. 2007). The sensitivity was higher for specimens containing more than 10^5 copies/ml of influenza virus RNA as determined by quantitative reverse transcription-polymerase chain reaction (qRT-PCR) (Cheng et al. 2009) or 10^3–10^5 $TCID_{50}$/ml of virus as determined by virus titration in cell cultures (Chan et al. 2009; Hurt et al. 2009). For avian samples, in which sensitivity of RT-PCR is known to be lower than that of virus isolation in embryonated chicken eggs, the sensitivity of antigen detection kits was comparable to that of RT-PCR (Cattoli et al. 2004); the minimum amount of virus needed was 5×10^4 $TCID_{50}$/ml (Fedorko et al. 2006).

2.2.2 Detection of Nucleic Acids

First described in 1985 (Saiki et al. 1985), the polymerase chain reaction (PCR) has been used to clone DNA, sequence, and analyze genes, identify people by their unique genetic fingerprint and diagnose infectious and genetic diseases. The production of complementary DNA (cDNA) from RNA was made possible by the development of RT-PCR. In 1992, PCR was made even more powerful with the innovation of real-time PCR (RRT-PCR) (Higuchi et al. 1992). Although semiquantitative in nature (Kubista et al. 2006), several RRT-PCR testing protocols have been developed for the detection and quantitation of influenza A viruses including SIVs (Spackman et al. 2002; Spackman and Suarez 2008).

The use of RNA extraction and purification methods varies by the type of sample being tested. For example, RNA can be extracted directly from infected amnioallantoic fluids, cell culture supernatants, bronchoalveolar lavage fluids (BALF), and oral fluids. However, for certain clinical diagnostic samples, prior processing is necessary. Tissue samples, such as lungs, are first made into a 10 % w/v homogenate using a balanced salt solution or a viral culture medium while nasal swabs are usually suspended and vortexed in a test tube with 2 ml of the above media. Although labor-intensive, standard organic extraction procedures produce high purity RNA from most any sample, including tissue homogenates, paraffin-embedded tissues, and body fluids (Sun 2010). However, commercial kits that use magnetic beads or solid-phase adsorption are more sensitive and easy to use with consistent results (Sun 2010). Commercial kits, such as RNeasy and QIAamp RNA kits (Qiagen, Valencia, CA) and PureLink™ RNA kit (Invitrogen, Carlsbad, CA) are based on solid-phase adsorption using silica-membrane spin columns. Commercial kits for magnetic bead extraction, such as MagMAX™ (Applied Biosystems, Foster City, CA) and EZ1 (Qiagen, Valencia, CA) are useful for liquid samples that have low virus concentration or contain PCR inhibitors, such as oral fluids, semen, urine, feces, and blood (Chan and McNally 2008; Das et al. 2009).

To detect a broad range of influenza A subtypes, primers for RRT-PCR are designed to target the conserved matrix (M) or nucleoprotein (NP) genes. The USDA-validated avian influenza RRT-PCR for the M gene (Spackman et al. 2002; Spackman and Suarez 2008) has been adapted for the detection of SIV in swine samples. The minimum detectable concentration of the virus for this procedure ranges from 10^{-1} to 10^{1} $TCID_{50}$/ml depending on the virus strain (Landolt et al. 2005; Richt et al. 2004). While virus isolation is still the gold standard test for influenza viruses, RT-PCR is an accurate, rapid, and sensitive technique that can be used to screen a large number of samples in a short period of time. The main disadvantage of RT-PCR is that it detects only the viral RNA and does not determine whether virus is viable or not. Since virus isolation depends on sample inoculation in a live culture system and detects the presence of live virus, it is often used in conjunction with RT-PCR to verify the presence of viable virus.

2.2.3 Detection of Whole Virus

Egg inoculation (EI) using nine to eleven-day-old embryonated chicken eggs is considered the gold standard for isolation and propagation of avian influenza viruses and certain egg-adapted SIVs (Clavijo et al. 2002; Swenson et al. 2001). However, it has been demonstrated that human influenza viruses propagated in chicken embryos acquired amino acid changes in their HA gene resulting in antigenic variation of the virus (Katz et al. 1987; Katz and Webster 1992; Meyer et al. 1993; Robertson et al. 1995). Comparatively, there was little to no genetic or antigenic variation in the same viruses when propagated in mammalian cell lines (Katz et al. 1987, 1990; Katz and Webster 1992; Meyer et al. 1993; Robertson et al. 1995), including Vero, MRC-5, BHK-21, and fetal porcine kidney cells. Of these, the Madin-Darby canine kidney (MDCK) cells have the highest sensitivity and are most commonly used in research and diagnostic applications (Meguro et al. 1979). For maximum sensitivity, inoculation of chicken embryos and/or another cell line is recommended in addition to MDCK cells.

Sample preparation for virus culture is the same as described for RT-PCR (Meguro et al. 1979). Influenza A viruses may replicate in cell cultures within 24–48 h or may take up to 5–6 days if the initial virus concentration in the sample is low. Growth of virus in cell cultures induces the production of cell lysis or cytopathic effects (CPE). Often a second blind passage is necessary for certain strains to show CPE. Once the virus has grown in cell cultures, tests can be performed on the culture supernatant to confirm viral identity. Although not a definitive assay, hemagglutination (HA) of chicken erythrocytes can be taken as a presumptive diagnosis of the virus and for approximation of the amount of virus present in the cell culture supernatant (1 HA unit approximates 5–6 $\log_{10}$ of virus). A more accurate method of quantifying virus is virus titration by inoculation of a set of serial dilutions in cell cultures (Villegas and Alvarado 2008). For definitive virus identification, the culture supernatant can be tested by RT-PCR or commercial influenza antigen test kits based on NP or M antigen. Since virus culture usually contains higher concentrations of virus than the original sample, sensitivity issue seen with clinical samples is usually not a problem when using antigen test kits.

Although virus isolation requires specialized equipment and maintenance of cell cultures and/or embryonated eggs, it is a standardized procedure that is available in most diagnostic laboratories. The virus isolated in cell culture can be cryogenically preserved for years and used for further characterization and vaccine production.

2.3 Indirect Detection

Although the clinical signs of influenza infection coincide with the presence of virus in nasal secretions, the isolation of virus by the gold standard method of virus culture or its detection by RT-PCR can be difficult when the period of virus shedding is brief. It has been found in vaccine challenge studies that shedding can be as transient as 24–72 h (Heinen et al. 2001, 2002; Van Reeth et al. 2001, 2003, 2006).

In situations when influenza virus is suspected but no longer detectable at the time of testing, detection of specific immunoglobulins may be undertaken. Immunoglobulins (predominantly IgG) are formed in swine at detectable levels within 1–2 weeks post infection and peak at 4–7 weeks (Olsen et al. 2006). For this reason, it has been recommended that serum samples be collected from pigs at the time of infection and at 3–4 weeks after the onset of clinical signs to compare the acute versus convalescent response (Rossow et al. 2003). Since influenza antibodies can be formed in response to both vaccination and exposure status, the interpretation of serologic assays will depend on both the vaccination and exposure status of the animals being tested. The serologic tests used to detect and measure influenza antibodies include: hemagglutination inhibition, serum neutralization, and enzyme-linked immunosorbent assays.

Hemagglutination inhibition (HI). The agglutination of red blood cells (RBCs) is a natural reaction that occurs in the presence of HA protein on the surface of the virus. HA can be specifically inhibited by influenza antibody, which can be measured in an HI assay. Optimum HA and HI reactions in SIVs occur with turkey or chicken RBCs, which are used in standardized tests (OIE 2008). Before conducting HI tests, it is imperative to remove non-specific inhibitors of viral hemagglutination and naturally occurring agglutinins from the serum samples to be tested. Inhibitors can be removed by treatment with receptor destroying enzyme (RDE) from *Vibrio cholerae,* heat inactivation, kaolin, or potassium periodate. Similarly, non-specific agglutinins can be removed by pretreatment of serum samples with chicken or turkey RBCs (Boliar et al. 2006; Pedersen 2008a; Regula et al. 2000; Ryan-Poirier and Kawaoka 1991; Springer and Ansell 1958; Subbarao et al. 1992). RDE and heat inactivation at 56 °C are the methods currently recommended to remove inhibitors (OIE 2008).

For the HI test, serial two fold dilutions of the test serum (starting at 1:10 and ending at 1:640 or 1:1280) are prepared in 96-well microtiter plates followed by the addition of 4–8 HA units of a single subtype of influenza virus in all wells containing serum dilutions. Following incubation for an hour at room temperature, 0.5 % suspension of RBCs is added to each well. In the absence of specific antibody, the virus is uninhibited (unbound) and is free to bind to the RBCs resulting in hemagglutination. However, if antihemagglutinin antibodies are present in the serum, such as after exposure or vaccination, the antibodies will bind to the hemagglutinin protein on the surface of the influenza virus, thus *inhibiting* the virus' ability to agglutinate the RBCs. The reciprocal of the highest serum dilution that inhibits HA is considered to be the HI titer of that serum (Fig. 2). HI titers greater than or equal to 1:40 are usually considered to be protective (Hancock et al. 2009).

The HI test is considered a standard test for the detection of SIV antibody (Villegas and Alvarado 2008) but is somewhat subjective in nature and the results may vary because of operator subjectivity and also upon repeating the test. Also, since there is broad cross-reactivity among the α, β, and γ clusters of the H1 subtype of SIVs, a positive HI titer may indicate a virus related to the virus of exposure, but does not definitively identify it. However, homologous virus reactions are typically stronger than heterologous virus reactions, resulting in higher HI titers. The advantages of this test are that it is a standardized procedure that is inexpensive and easy to perform and the results are comparable to more complicated tests, such as serum neutralization (Leuwerke et al. 2008; Vincent et al. 2006, 2009a).

Serum neutralization (SN) or virus neutralization (VN). The SN test detects virus-specific neutralizing antibody present in a serum sample. Serial two fold dilutions of the serum and a known amount of SIV are preincubated and then

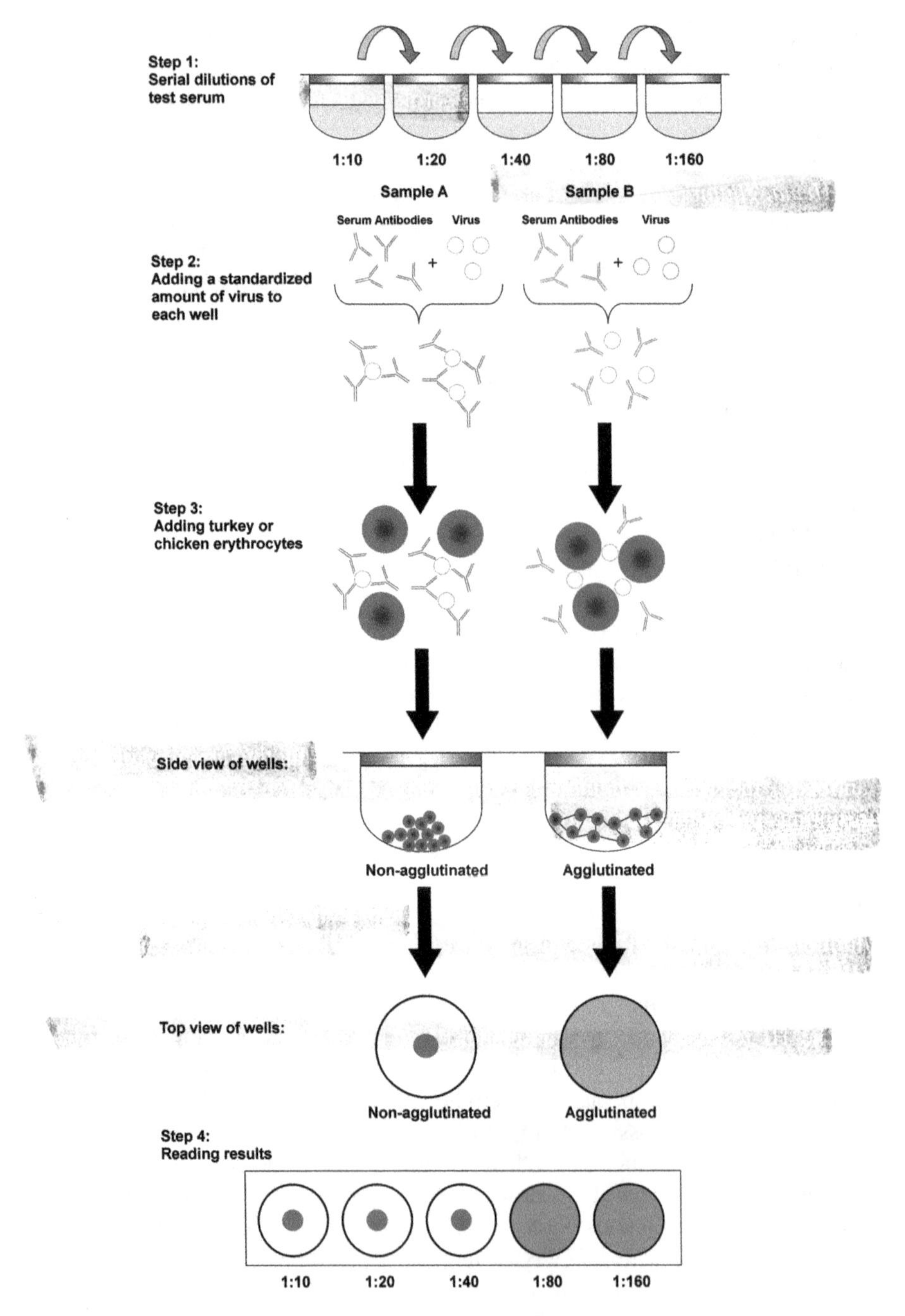

Fig. 3 Steps in a hemagglutination inhibition reaction. The antibodies on the left in sample A prevent the virus from agglutinating the erythrocytes. Whereas the antibodies on the right in sample B do not bind to the virus in step 2, which agglutinate the erythrocytes in step 3. The antibody titer shown in step 4 is read out as 1:40

added to MDCK cells to determine the highest dilution of serum that can neutralize virus infection of cells and production of CPE (Fig. 3). Neutralizing antibodies in serum sample block viral infection of cell culture and the virus is not available to produce CPE. However, if antibodies are not present, the virus is not blocked and is free to cause CPE in inoculated cell cultures. Reciprocal of the highest serum dilution that can neutralize virus infection is considered to be the SN titer of the serum. Since the test uses very small volumes of serum in cell monolayers contained in 96-well microtiter plates, it is often called micro neutralization. One of the advantages of SN over HI and enzyme-linked immunosorbent assays (ELISA) is that it demonstrates the biologic (neutralizing) activity of the antibodies present in the serum. Some of the disadvantages of this test are that it requires equipment and supplies used for virus cultures and the results can take up to 72 h to obtain. Also, the SN titers may vary when the test is repeated.

Enzyme-linked immunosorbent assay (ELISA). The ELISA test uses a 96-well plate that has been coated with influenza viral antigen. The serum sample is incubated in the coated wells for antibody attachment. After the unbound material is washed away, an anti-influenza monoclonal antibody that is conjugated to an enzyme is bound to the antigen. The unbound conjugate is washed away and the enzyme substrate (that produces a color change in the presence of the enzyme) is added to the wells. The color-changing reaction is stopped after 15 min and the amount of color produced is read as an optical density (O.D.) in a spectrophotometer (Fig. 4). The O.D. is inversely proportional to the amount of anti-influenza antibodies present in the test sample. Commercially available ELISA test kits include separate ELISA tests for H1N1 and H3N2 subtypes of SIV. Another ELISA that detects antibodies to a range of influenza A viruses is available and has been adapted for use in detecting anti-SIV antibodies (Ciacci-Zanella et al. 2010). The commercial H1N1 ELISA uses an antigen prepared from a classical H1N1 SIV and, thus has a limited detection range of swine H1 subtypes. Although the H1N1 test is not designed to detect other influenza subtypes, it may sometimes cross-react with H3N2 because of some common epitopes between H1N1 and H3N2 viruses. In addition, the H1N1 test has been found to miss recently infected animals (Yoon et al. 2004). The H3N2 ELISA test was developed from a cluster I virus leading to lower reactivity with class IV viruses (Yoon et al. 2004). The MultiS-Screen ELISA (FlockChek™, Idexx, Westbrook, ME) uses a highly conserved epitope of influenza A nucleoprotein (NP) (Ciacci-Zanella et al. 2010). Preliminary studies indicate that this kit, while originally designed for use in avian species, also detects antibodies against subtypes common to swine (Ciacci-Zanella et al. 2010) (Fig. 5).

2.4 Virus Subtyping and Sequencing

Important for host range, antigenicity, and pathogenesis, the 16 HA and 9 NA genes are antigenically and genetically divergent and these variations are used for subtyping the influenza viruses. The cultured viruses were traditionally subtyped

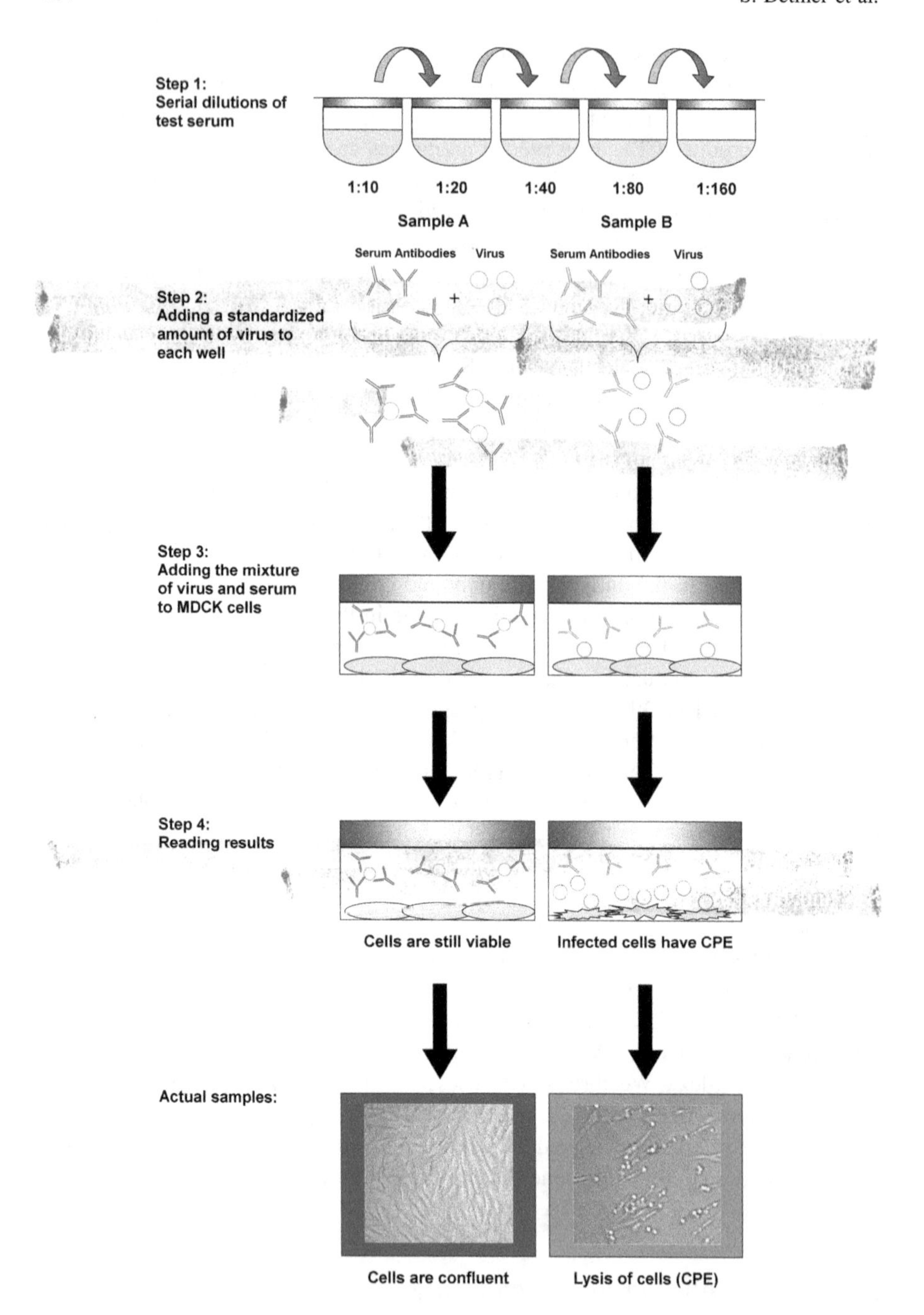

Fig. 4 Steps in a serum neutralization reaction. The antibodies in sample A on the left neutralized the virus in step 2. This resulted in no cytopathic effects (CPE) in step 4. Whereas the antibodies in sample B on the right did not neutralize the virus in step 2, resulting in infection of the MDCK cells and CPE in step 4

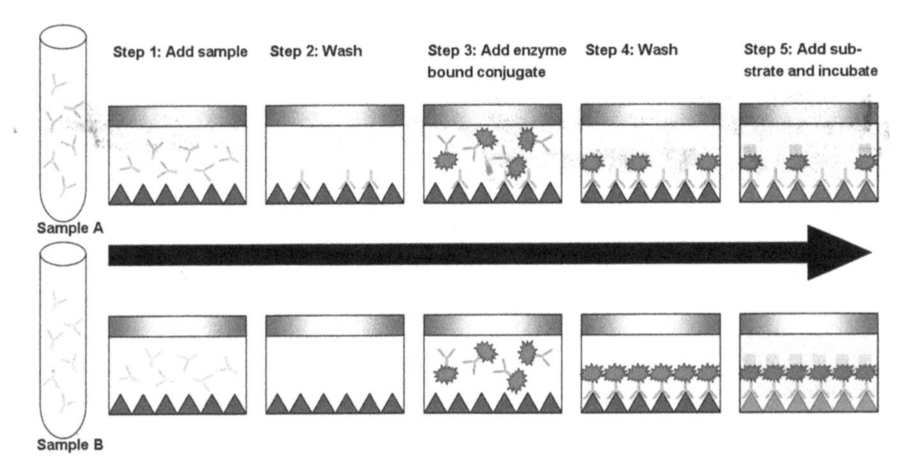

Fig. 5 Steps in a blocking ELISA test. The optical density of sample A is lower than sample B because the influenza A antibody in sample A is bound to the antigen coated on the bottom of the well, partially blocking the binding of the enzyme bound conjugate. The antibodies in sample B did not bind to the antigen and were therefore washed out in step 2. Figure adapted from http://www.idexx.com/pubwebresources/pdf/en_us/livestock-poultry/0965846.pdf

using HI and NA inhibition (NI) assays (Pedersen 2008a, b). The NI assay uses a dilution of the cultured virus between 1:4 and 1:32, depending on the virus concentration. There are several steps that include standardized NA antisera (N1–N9), fetuin, periodate, sodium arsenite, and thiobarbituric acid which result in a dark color if there is no inhibition and a light color if there is inhibition; the NA subtype has the light color result. Both of these assays are time-consuming and require standardized NA and HA antisera, which are often difficult to acquire. Therefore, RT-PCR is now regularly used for subtyping. Currently, HA and NA specific primers can be used for both detection and subtyping of influenza A viruses. Additionally, a number of multiplex and nested RT-PCR have been developed for subtyping with and without simultaneous detection of influenza A virus (Chander et al. 2010; Fereidouni et al. 2009; He et al. 2009; Lam et al. 2007; Li et al. 2001; Stockton et al. 1998; Yang et al. 2010).

In addition to subtyping, RT-PCR can also be used for sequencing all eight gene segments of influenza virus (Chander et al. 2010; Jindal et al. 2009). The sequences can be examined and compared to other sequences with molecular analysis tools; uncovering the evolutionary; and geographic relationships of influenza viruses. However, the amount of RNA in clinical samples is usually low compared to the other cellular materials and contaminating bacteria (Spackman and Suarez 2008). Therefore, cell culture supernatants and amnio-allantoic fluid containing a large concentration of whole virus, are recommended for sequencing and other molecular analyses (Spackman and Suarez 2008).

2.5 Limitations of Diagnostic Assays

The rapid evolution of influenza A viruses over the last decade has led to genetic and antigenic variation of the virus in North American swine. This has led to limitations in cross-reactivity for the serologic assays. These changes need to be kept in mind when interpreting the results of these tests. Although there is some antigenic cross-reactivity among the classical and reassorted α, β, and γ clusters of the swine H1 subtype, there is little to no cross-reactivity between these three clusters and the human-like δ cluster (Vincent et al. 2006, 2009b). This variability in the antigenic cross-reactivity was demonstrated in 2009 pandemic H1N1 virus for both North American and European swine H1 subtypes using sera from experimentally infected and vaccinated pigs (Kyriakis et al. 2010b; Vincent et al. 2010). The human-like viruses in the SwH1δ cluster were recently found to have two distinct antigenically divergent groups, which could result in additional limitations for serologic assays (Vincent et al. 2009a). Similarly among the swine H3 viruses, there is little to no cross-reactivity between groups I and IV. There is also limited to no cross-reactivity between swine subtypes, which means that multiple viruses from each subtype need to be tested to determine the subtype of the virus that produced the antibodies. To overcome the limitations of cross-reactivity and broaden influenza surveillance, the samples may first be screened by the MultiS-Screen ELISA followed by more specific tests, such as SN and HI assays to determine the subtype of the virus of exposure.

As the influenza virus continues to evolve, the primers for RT-PCR for detection and subtyping need to be continually validated and updated. Current testing stratagems rely on conserved nucleotide sequences for the primers. However, the variability in the HA and NA genes in avian influenzas have resulted in the design of multiple *wobble primers* to detect one subtype of influenza A without cross-reactivity with other HA and NA subtypes (Sidoti et al. 2010; Starick et al. 2000; Suarez et al. 2007). The avian influenza primers can be used for subtyping influenza viruses from swine or new subtyping primers can be designed using published sequences (He et al. 2009; Huang et al. 2009; Lee et al. 2008; Nagarajan et al. 2010). New technologies, such as enzyme hybridization and microarray, are being used for subtyping of influenza viruses across species (avian, human and swine) and detection of specific influenza viruses like 2009 pandemic H1N1 (He et al. 2009; Huang et al. 2009).

References

AASV [American Association of Swine Veterinarians] (2009) American Association of Swine Veterinarians position statement on pandemic (H1N1) 2009 influenza. www.aasv.org/aasv/position-pH1N1.pdf. Accessed 28 May 2010

AASV [American Association of Swine Veterinarians] (2010) Comprehensive and integrated swine surveillance. http://www.aasv.org/shap/issues/v18n2/v18n2advocacy.htm. Accessed 25 June 2010

Boliar S, Stanislawek W, Chambers TM (2006) Inability of kaolin treatment to remove nonspecific inhibitors from equine serum for the hemagglutination inhibition test against equine H7N7 influenza virus. J Vet Diagn Invest 18:264–267

Bos MEH, te Beest DE, van Boven M, Robert-Du Ry van Beest Holle M, Meijer A, Bosman A, Mulder YM, Koopmans MPG, Stegeman A (2010) High probability of avian influenza virus (H7N7) transmission from poultry to humans ctive in disease control on infected farms. J Infect Dis 201:1390–1396

Brockmeier S, Halbur PG, Thacker EL (2002) Porcine respiratory disease complex. In: Brogden K, Guthmiller JM (eds) Polymicrobial diseases. ASM Press, Washington , pp 231–257

Brown IH (2000) The epidemiology and evolution of influenza viruses in pigs. Vet Microbiol 74:29–46

Buchwalow IB, Böcker W, SpringerLink (2010) Immunohistochemistry: basics and methods. Springer, Berlin

Cannon RM, Roe RT (1982) Livestock disease surveys: a field manual for veterinarians. Australian Government Publishing Services, Canberra, pp 14

Cattoli G, Drago A, Maniero S, Toffan A et al (2004) Comparison of three rapid detection systems for type A influenza virus on tracheal swabs of experimentally and naturally infected birds. Avian Pathol 33:432–437

CDC [Centers for Disease Control and Prevention] (2009) Lesson 5: public health surveillance. In: principles of epidemiology in public health practice, 3rd ed. http://www.cdc.gov/training/products/ss1000/ss1000-ol.pdf Accessed 23 April 2012

Chambers TM, Hinshaw VS, Kawaoka Y, Easterday BC, Webster RG (1991) Influenza viral infection of swine in the United States 1988–1989. Arch Virol 116:261–265

Chan DJ, McNally L (2008) Assays for the determination of HIV-1 load in semen: a review of indications, methods and performance in vitro. Current HIV Res 6:182–188

Chan KH, Lai ST, Poon LL, Guan Y, Yuen K, Peiris JS (2009) Analytical sensitivity of rapid influenza antigen detection tests for swine-origin influenza virus (H1N1). J Clin Virol 45:205–207

Chander Y, Jindal N, Stallknecht D et al (2010) Full length sequencing of all nine subtypes of the neuraminidase gene of influenza A viruses by primer walking. J Virol Meth 165:116–120

Cheng CK, Cowling BJ, Chan KH, Fang VJ et al (2009) Factors affecting quickvue Influenza A + B rapid test performance in the community setting. Diagn Microbiol Infect Dis 65:35–41

Ciacci-Zanella JR, Vincent AL, Prickett JR, Zimmerman SM, Zimmerman JJ (2010) Detection of anti-influenza A nucleoprotein antibodies in pigs using a commercial influenza epitope-blocking enzyme-linked immunosorbent assay developed for avian species. J Vet Diagn Invest 22:3–9

Clavijo A, Tresnan DB, Jolie R, Zhou EM (2002) Comparison of embryonated chicken eggs with MDCK cell culture for the isolation of swine influenza virus. Canadian J Vet Res 66:117–121

Das A, Spackman E, Pantin-Jackwood MJ, Suarez DL (2009) Removal of real-time reverse transcription polymerase chain reaction (RT-PCR) inhibitors associated with cloacal swab samples and tissues for improved diagnosis of avian influenza virus by RT-PCR. J Vet Diagn Invest 21:771–778

Davies PR, Wayne SR, Torrison JL, Peele B, de Groot BD, Wray D (2007) Real-time disease surveillance tools for the swine industry in Minnesota. Vet Ital 43:731–738

Domenech J, Lubroth J, Eddi C, Martin V, Roger F (2006) Regional and international approaches on prevention and control of animal transboundary and emerging diseases. Ann NY Acad Sci 1081:90–107

Elbers AR, Vos JH, Bouma A, van Exsel AC, Stegeman A (2003) Assessment of the use of gross lesions at post-mortem to detect outbreaks of classical swine fever. Vet Microbiol 96:345–356

Engel B, Bouma A, Stegeman A, Buist W, Elbers A, Kogut J, Döpfer D, de Jong MC (2005) When can a veterinarian be expected to detect classical swine fever virus among breeding sows in a herd during an outbreak? Prev Vet Med 67(2–3):195–212

ESNIP (2007) The European surveillance network for influenza in pigs. http://www.esnip.ugent.be. Accessed 20 June 2010

European Commission (2010) European Commission unveils new research projects to fight influenza. http://ec.europa.eu/research/index.cfm?pg=newsalert&lg=en&year=2010&na=na-090310-annexes. Accessed 21 June 2010

Fedorko DP, Nelson NA, McAuliffe JM, Subbarao K (2006) Performance of rapid tests for detection of avian influenza A virus types H5N1 and H9N2. J Clin Microbiol 44:1596–1597
Fereidouni SR, Starick E, Grund C, Globig A et al (2009) Rapid molecular subtyping by reverse transcription polymerase chain reaction of the neuraminidase gene of avian influenza A viruses. Vet Microbiol 135(3–4):253–260
Gramer MR, Torrison JL (2010) Happy anniversary to (pandemic 2009 H1N1) flu. National Hog Farmer Weekly Preview. http://nationalhogfarmer.com/weekly-preview/0503-happy-anniversary-pandemic-2009-h1n1/. Accessed 23 June 2010
Guarner J, Shieh WJ, Dawson J, Subbarao K et al (2000) Immunohistochemical and in situ hybridization studies of influenza A virus infection in human lungs. Am J Clin Pathol 114: 227–233
Gubernot DM, Boyer BL, Moses MS (2008) Animals as early detectors of bioevents: veterinary tools and a framework for animal-human integrated zoonotic disease surveillance. Public Health Rep 123:300–315
Haines DM, Waters EH, Clark EG (1993) Immunohistochemical detection of swine influenza A virus in formalin-fixed and paraffin-embedded tissues. Canadian J Vet Res 57:33–36
Hancock K, Veguilla V, Lu X, Zhong W et al (2009) Cross-reactive antibody responses to the 2009 pandemic H1N1 influenza virus. N Engl J Med 361:1945–1952
He J, Bose ME, Beck ET, Fan J et al (2009) Rapid multiplex reverse transcription-PCR typing of influenza A and B virus, and subtyping of influenza A virus into H1, 2, 3, 5, 7, 9, N1 (human), N1 (animal), N2, and N7, including typing of novel swine origin influenza A (H1N1) virus, during the 2009 outbreak in Milwaukee, Wisconsin. J Clin Microbiol 47:2772–2778
Heinen PP, van Nieuwstadt AP, de Boer-Luijtze EA, Bianchi AT (2001) Analysis of the quality of protection induced by a porcine influenza A vaccine to challenge with an H3N2 virus. Vet Immunol Immunopathol 82:39–56
Heinen PP, Rijsewijk FA, de Boer-Luijtze EA, Bianchi AT (2002) Vaccination of pigs with a DNA construct expressing an influenza virus M2-nucleoprotein fusion protein exacerbates disease after challenge with influenza A virus. J Gen Virol 83:1851–1859
Hermann JR, Hoff SJ, Yoon KJ, Burkhardt AC, Evans RB, Zimmerman JJ (2006) Optimization of a sampling system for recovery and detection of airborne porcine reproductive and respiratory syndrome virus and swine influenza virus. Appl Environ Microbiol 72:4811–4818
Higuchi R, Dollinger G, Walsh PS, Griffith R (1992) Simultaneous amplification and detection of specific DNA sequences. Biotechnology (NY) 10:413–417
Hinshaw VS, Bean WJ, Webster RG, Easterday BC (1978) The prevalence of influenza viruses in swine and the antigenic and genetic relatedness of influenza viruses from man and swine. Virology 84:51–62
Hofshagen M, Gjerset B, Er C, Tarpai A, Brun E, Dannevig B, Bruheim T, Fostad IG, Iversen B, Hungnes O, Lium B (2009) Pandemic influenza A(H1N1)v: human to pig transmission in Norway? Euro Surveill 14, ii:19406
Huang Y, Tang H, Duffy S, Hong Y et al (2009) Multiplex assay for simultaneously typing and subtyping influenza viruses by use of an electronic microarray. J Clin Microbiol 47:390–396
Hurt AC, Alexander R, Hibbert J, Deed N, Barr IG (2007) Performance of six influenza rapid tests in detecting human influenza in clinical specimens. J Clin Virol 39:132–135
Hurt AC, Baas C, Deng YM, Roberts S, Kelso A, Barr IG (2009) Performance of inflenza rapid point-of-care tests in the detection of swine lineage A(H1N1) influenza viruses. Influenza Other Respir Viruses 3:171–176
IOM/NRC [National Research Council] (2009) Sustaining global surveillance and response to emerging zoonotic diseases. National Academies Press, Washington
Jajosky R, Groseclose S (2004) Evaluation of reporting timeliness of public health surveillance systems for infectious diseases. BMC Public Health 4:29
Jindal N, Chander Y, Sreevatsan S et al (2009) Amplification of four different genes of influenza A viruses using a degenerate primer set in a one step RT-PCR method. J Virol Methods 160:163–166

Karasin AI, Olsen CW, Brown IH, Carman S, Stalker M (2000) H4N6 influenza virus isolated from pigs in Ontario. Can Vet J 41:938–939

Katz JM, Webster RG (1992) Amino acid sequence identity between the HA1 of influenza A (H3N2) viruses grown in mammalian and primary chick kidney cells. J Gen Virol 73: 1159–1165

Katz JM, Naeve CW, Webster RG (1987) Host cell-mediated variation in H3N2 influenza viruses. Virology 156:386–395

Katz JM, Wang M, Webster RG (1990) Direct sequencing of the HA gene of influenza (H3N2) virus in original clinical samples reveals sequence identity with mammalian cell-grown virus. J Virol 64:1808–1811

Kida H, Ito T, Yasuda J, Shimizu Y, Itakura C, Shortridge KF, Kawaoka Y, Webster RG (1994) Potential for transmission of avian influenza viruses to pigs. J Gen Virol 75:2183–2188

Kubista M, Andrade JM, Bengtsson M, Forootan A et al (2006) The real-time polymerase chain reaction. Mol Aspects Med 27:95–125

Kyriakis CS, Brown IH, Foni E, Kuntz-Simon G, Maldonado J, Madec F, Essen SC, Chiapponi C, Van Reeth K (2010a) Virological surveillance and preliminary antigenic characterization of influenza viruses in pigs in five European countries from 2006 to 2008. Zoonoses Pub H. doi:10.1111/j.1863-2378.2009.01301.x

Kyriakis CS, Olsen CW, Carman S, Brown IH et al (2010b) Serologic cross-reactivity with pandemic (H1N1) 2009 virus in pigs, Europe. Emerg Infect Dis 16:96–99

Lam WY, Yeung AC, Tang JW, Ip M et al (2007) Rapid multiplex nested PCR for detection of respiratory viruses. J Clin Microbiol 45:3631–3640

Landolt GA, Karasin AI, Hofer C, Mahaney J, Svaren J, Olsen CW (2005) Use of real-time reverse transcriptase polymerase chain reaction assay and cell culture methods for detection of swine influenza A viruses. Am J Vet Res 66:119–124

Larochelle R, Sauvageau R, Magar R (1994) Immunohistochemical detection of swine influenza virus and porcine reproductive and respiratory syndrome virus in porcine proliferative and necrotizing pneumonia cases from Quebec. Canadian Vet J 35:513–515

Lee CS, Kang BK, Lee DH, Lyou SH et al (2008) One-step multiplex RT-PCR for detection and subtyping of swine influenza H1, H3, N1, N2 viruses in clinical samples using a dual priming oligonucleotide (DPO) system. J Virol Methods 151:30–34

Leuwerke B, Kitikoon P, Evans R, Thacker E (2008) Comparison of three serological assays to determine the cross-reactivity of antibodies from eight genetically diverse U.S. swine influenza viruses. J Vet Diagn Invest 20:426–432

Li J, Chen S, Evans DH (2001) Typing and subtyping influenza virus using DNA microarrays and multiplex reverse transcriptase PCR. J Clin Microbiol 39:696–704

Lynn T, Marano N, Treadwell T, Bokma B (2006) Linking human and animal health surveillance for emerging diseases in the United States achievements and challenges. Ann NY Acad Sci 1081:108–111

Ma W, Gramer M, Rossow K, Yoon KJ (2006) Isolation and genetic characterization of new reassortant H3N1 swine influenza virus from pigs in the midwestern United States. J Virol 80:5092–5096

Meguro H, Bryant JD, Torrence AE, Wright PF (1979) Canine kidney cell line for isolation of respiratory viruses. J Clin Microbiol 9:175–179

Meyer WJ, Wood JM, Major D, Robertson JS, Webster RG, Katz JM (1993) Influence of host cell-mediated variation on the international surveillance of influenza A (H3N2) viruses. Virology 196:130–137

Myers KP, Olsen C, Setterquist SF, Capuano AW, Donham KJ, Thacker EL, Merchant JA, Gray GC (2006) Are swine workers in the United States at increased risk of infection with zoonotic influenza virus? Clin Infect Dis 42:14–20

Nagarajan MM, Simard G, Longtin D, Simard C (2010) Single-step multiplex conventional and real-time reverse transcription polymerase chain reaction assays for simultaneous detection and subtype differentiation of Influenza A virus in swine. J Vet Diagn Invest 22:402–408

NIAID [National Institute of Allergy and Infectious Diseases] (2010) Centers of Excellence for Influenza Research and Surveillance (CEIRS). http://www.niaid.nih.gov/LabsAndResources/resources/ceirs/Pages/introduction.aspx. Accessed 25 June 2010

OIE [Office International des Epizooties] (2009) Pandemic H1N1: questions and answers. www.oie.int/eng/press/h1n1/en_h1_n1_faq.asp. Accessed 29 May 2010

OIE [World Organization for Animal Health] (2008) Manual of diagnostic tests and vaccines for terrestrial animals. Chapter 2.8.8: Swine Influenza, http://www.oie.int/fr/normes/mmanual/2008/pdf/2.08.08_SWINE_INFLUENZA.pdf. Accessed 22 June 2010

Olsen CW, Carey S, Hinshaw L, Karasin AI (2000) Virologic and serologic surveillance for human, swine and avian influenza virus infections among pigs in the north-central United States. Arch Virol 145:1399–1419

Olsen CW, Brown IH, Easterday BC, Van Reeth K (2006) Swine Influenza. In: Straw BE et al (eds) Diseases of swine, 9th edn. Blackwell, Ames, pp 469–482

Onno M, Jestin A, Vannier P, Kaiser C (1990) Diagnosis of swine influenza with an immunofluorescence technique using monoclonal antibodies. Vet Q 12:251–254

Pappaioanou M, Gramer M (2010) Lessons from pandemic H1N1 2009 to improve prevention, detection, and response to influenza pandemics from a one health perspective. ILAR J 51: 268–280

Patriarca PA, Cox NJ (1997) Influenza pandemic preparedness plan for the United States. J Infect Dis 176:S4–S7

Pedersen JC (2008a) Hemagglutination-inhibition test for avian influenza virus subtype identification and the detection and quantitation of serum antibodies to the avian influenza virus. In: Spackman E (ed) Methods in Molecular biology, Avian influenza virus, vol 436. Humana Press, Totowa

Pedersen JC (2008b) Neuraminidase-inhibition assay for the identification of influenza A virus neuraminidase subtype or neuraminidase antibody specificity. In: Spackman E (ed) Methods in Molecular biology, Avian influenza virus, vol 436. Humana Press, Totowa

Peiris JSM, Guan Y et al (2001) Cocirculation of avian H9N2 and contemporary "human" H3N2 influenza A viruses in pigs in southeastern China: Potential for genetic reassortment? J Virol 75:9679–9686

Pensaert M, Ortis K, Vandeputte J, Kaplan MM, Bachmann PA (1981) Evidence for the natural transmission of influenza A virus from wild ducks to swine and its potential importance for man. Bull World Hlth Org 59:75–78

Poliak Z, Dewey CE, Martin SW, Christensen J, Carman S, Friendship RM (2008) Prevalence of and risk factors for influenza in southern Ontario swine herds in 2001 and 2003. Can J Vet Res 72:7–17

Prickett JR, Zimmerman JJ (2010) The development of oral fluid-based diagnostics and applications in veterinary medicine. Animal Health Res Rev 5:1–10

Radostits OM, Gay CC, Blood DC, Hinchcliff KW (2000) Veterinary medicine: a textbook of the diseases of cattle, sheep, pigs, goats, and horses, 9th edn. WB Saunders Co, New York, pp 1157–1159

Regula G, Lichtensteiger CA, Mateus-Pinilla NE, Scherba G, Miller GY, Weigel RM (2000) Comparison of serologic testing and slaughter evaluation for assessing the effects of subclinical infection on growth in pigs. J Am Vet Med Assoc 217:888–895

Richt JA, Lager KM, Clouser DF, Spackman E, Suarez DL, Yoon KJ (2004) Real-time reverse transcription-polymerase chain reaction assays for the detection and differentiation of North American swine influenza viruses. J Vet Diagn Invest 16:367–373

Richt JA, Lekcharoensuk P, Lager KM, Vincent AL et al (2006) Vaccination of pigs against swine influenza viruses by using an NS1-truncated modified live-virus vaccine. J Virol 80:11009–11018

Robertson JS, Cook P, Attwell A, Williams SP (1995) Replicative advantage in tissue culture of egg-adapted influenza virus over tissue-culture derived virus: implications for vaccine manufacture. Vaccine 13:1583–1588

Rossow KD, Yeske P, Goyal SM, Webby RJ, Collins JE (2003) Diagnostic investigation of unexpected serology results for swine influenza virus (SIV) and porcine reproductive and respiratory syndrome virus (PRRSV). J Swine Hlth Prod 1133–1135
Ryan-Poirier KA, Kawaoka Y (1991) Distinct glycoprotein inhibitors of influenza A virus in different animal sera. J Virol 65:389–395
Sabattini E, Bisgaard K, Ascani S, Poggi S et al (1998) The EnVision++ system: a new immunohistochemical method for diagnostics and research. Critical comparison with the APAAP, ChemMate, CSA, LABC, and SABC techniques. J Clin Pathol 51:506–511
Saiki RK, Scharf S, Faloona F, Mullis KB et al (1985) Enzymatic amplification of beta-globin genomic sequences and restriction site analysis for diagnosis of sickle cell anemia. Science 230:1350–1354
Schnurrenberger P, Sharman RS, Wise GH (1987) Attacking animal diseases: concepts and strategies for control and eradication. Iowa State University Press, Ames, p 64
Shaman J, Kohn M (2009) Absolute humidity modulates influenza survival, transmission, and seasonality. Proc Natn Acad Sci 106:3243–3248
Sidoti F, Rizzo F, Costa C, Astegiano S et al (2010) Development of real time RT-PCR assays for detection of type A influenza virus and for subtyping of avian H5 and H7 hemagglutinin subtypes. Mol Biotechnol 44:41–50
Smith GJD, Vijaykrishna D et al (2009) Origins and evolutionary genomics of the 2009 swine origin H1N1 influenza A epidemic. Nature 459:1122–1125
Spackman E, Suarez DL (2008) Type A influenza virus detection and quantitation by real-time RT-PCR. In: Spackman E (ed) Methods in molecular biology, Avian influenza virus, vol 436. Humana Press, Totowa
Spackman E, Senne DA, Myers TJ, Bulaga LL et al (2002) Development of a real-time reverse transcriptase PCR assay for type A influenza virus and the avian H5 and H7 hemagglutinin subtypes. J Clin Microbiol 40:3256–3260
Springer GF, Ansell NJ (1958) Inactivation of human erythrocyte agglutinogens M and N by influenza viruses and receptor-destroying enzyme. Proc Natl Acad Sci USA 44:182–189
Starick E, Romer-Oberdorfer A, Werner O (2000) Type- and subtype-specific RT-PCR assays for avian influenza A viruses (AIV). J Vet Med B Infect Dis Vet Public Health 47:295–301
Stockton J, Ellis JS, Saville M, Clewley JP, Zambon MC (1998) Multiplex PCR for typing and subtyping influenza and respiratory syncytial viruses. J Clin Microbiol 36:2990–2995
Straw BE, Dewey CE, Wilson MR (2006) Differential diagnosis of disease. In: Straw BE et al (eds) Diseases of swine, 9th edn. Blackwell, Ames, pp 241–286
Suarez DL, Das A, Ellis E (2007) Review of rapid molecular diagnostic tools for avian influenza virus. Avian Dis 51:201–208
Subbarao EK, Kawaoka Y, Ryan-Poirier K, Clements ML, Murphy BR (1992) Comparison of different approaches to measuring influenza a virus-specific hemagglutination inhibition antibodies in the presence of serum inhibitors. J Clin Microbiol 30:996–999
Sun W (2010) Nucleic extraction and amplification. In: Grody W et al (eds) Molecular diagnostics: techniques and applications for the clinical laboratory, 1st edn. Academic Press, San Diego, pp 35–47
Swenson SL, Vincent LL, Lute BM, Janke BH et al (2001) A comparison of diagnostic assays for the detection of type A swine influenza virus from nasal swabs and lungs. J Vet Diagn Invest 13:36–42
Torremorell M, Juarez A, Chavez E, Yescas J, Doporto JM, Gramer M (2009) Procedures to eliminate H3N2 swine influenza virus from a pig herd. Vet Rec 165:74–77
USDA/APHIS [Animal and Plant Health Inspection Service] (2009) National Surveillance Plan for Swine Influenza Virus: including novel H1N1 2009 virus. www.aphis.usda.gov/newsroom/hot_issues/h1n1/downloads/H1N1_Surveillance_Plan_2009.pdf. Accessed 21 Dec 2009
Van Reeth K, Nauwynck H, Pensaert M (1996) Dual infections of feeder pigs with porcine reproductive and respiratory syndrome virus followed by porcine respiratory coronavirus or swine influenza virus: a clinical and virological study. Vet Microbiol 48:325–335

Van Reeth K, Labarque G, De Clercq S, Pensaert M (2001) Efficacy of vaccination of pigs with different H1N1 swine influenza viruses using a recent challenge strain and different parameters of protection. Vaccine 19:4479–4486

Van Reeth K, Gregory V, Hay A, Pensaert M (2003) Protection against a European H1N2 swine influenza virus in pigs previously infected with H1N1 and/or H3N2 subtypes. Vaccine 21:1375–1381

Van Reeth K, Labarque G, Pensaert M (2006) Serological profiles after consecutive experimental infections of pigs with European H1N1, H3N2, and H1N2 swine influenza viruses. Viral Immunol 19:373–382

Vijaykrishna D, Poon LLM, Zhu HC et al (2010) Reassortment of pandemic H1N1/2009 influenza A virus in swine. Science 328:1529

Villegas P, Alvarado I (2008) Chapter 46. In: Dufour-Zavala L, et al. (ed) A laboratory manual for the isolation, identification, and characterization of avian pathogens 5th edn. American Association of Avian Pathologists, Jacksonville, Florida, p 218

Vincent LL, Janke BH, Paul PS, Halbur PG (1997) A monoclonal-antibody-based immunohistochemical method for the detection of swine influenza virus in formalin-fixed, paraffin-embedded tissues. J Vet Diagn Invest 9:191–195

Vincent AL, Lager KM, Ma W, Lekcharoensuk P et al (2006) Evaluation of hemagglutinin subtype 1 swine influenza viruses from the United States. Vet Microbiol 118:212–222

Vincent AL, Ma W, Lager KM, Janke BH, Richt JA (2008) Swine influenza viruses a North American perspective. Adv Virus Res 72:127–154

Vincent AL, Lager KM, Harland M, Lorusso A, Zanella E et al (2009a) Absence of 2009 pandemic H1N1 Influenza A virus in fresh pork. PLoS ONE 4:e8367. doi:10.1371/journal.pone.0008367

Vincent AL, Ma W, Lager KM, Gramer MR, Richt JA, Janke BH (2009b) Characterization of a newly emerged genetic cluster of H1N1 and H1N2 swine influenza virus in the United States. Virus Genes 39:176–185

Vincent AL, Lager KM, Faaberg KS, Harland M et al (2010) Experimental inoculation of pigs with pandemic H1N1 2009 virus and HI cross-reactivity with contemporary swine influenza virus antisera. Influenza other Respir Viruses 4:53–60

Vosse BA, Seelentag W, Bachmann A, Bosman FT, Yan P (2007) Background staining of visualization systems in immunohistochemistry: comparison of the Avidin-Biotin Complex system and the EnVision+ system. Appl Immunohistochem Mol Morphol 15:103–107

Webby RJ, Rossow K, Erickson G, Sims Y, Webster R (2004) Multiple lineages of antigenically and genetically diverse influenza A virus co-circulate in the United States swine population. Virus Res 103:67–73

Yang Y, Gonzalez R, Huang F, Wang W et al (2010) Simultaneous typing and HA/NA subtyping of influenza A and B viruses including the pandemic influenza A/H1N1 2009 by multiplex real-time RT-PCR. J Virol Methods 167:37–44

Yazawa S, Okada M, Ono M et al (2004) Experimental dual infection of pigs with an H1N1 swine influenza virus (A/Sw/Hok/2/81) and *Mycoplasma hyopneumoniae*. Vet Microbiol 98:221–228

Yoon KJ, Janke BH, Swalla RW, Erickson G (2004) Comparison of a commercial H1N1 enzyme-linked immunosorbent assay and hemagglutination inhibition test in detecting serum antibody against swine influenza viruses. J Vet Diagn Invest 16:197–201

Contemporary Epidemiology of North American Lineage Triple Reassortant Influenza A Viruses in Pigs

Alessio Lorusso, Amy L. Vincent, Marie R. Gramer, Kelly M. Lager and Janice R. Ciacci-Zanella

Abstract The 2009 pandemic H1N1 infection in humans has been one of the greatest concerns for public health in recent years. However, influenza in pigs is a zoonotic viral disease well-known to virologists for almost one century with the classical H1N1 subtype the only responsible agent for swine influenza in the United States for many decades. Swine influenza was first recognized clinically in pigs in the Midwestern U.S. in 1918 and since that time it has remained important to the swine industry throughout the world. Since 1998, however, the epidemiology of swine influenza changed dramatically. A number of emerging subtypes and genotypes have become established in the U.S. swine population. The ability of multiple influenza virus lineages to infect pigs is associated with the emergence of reassortant viruses with new genomic arrangements, and the introduction of the 2009 pandemic H1N1 from humans to swine represents a well-known example. The recent epidemiological data regarding the current state of influenza A virus subtypes circulating in the Canadian and American swine population is discussed in this review.

A. Lorusso · A. L. Vincent (✉) · K. M. Lager · J. R. Ciacci-Zanella
Virus and Prion Diseases Research Unit, National Animal Disease Center, USDA, Agricultural Research Service, Ames, IA, USA
e-mail: amy.vincent@ars.usda.gov

M. R. Gramer
University of Minnesota, St. Paul, MN, USA

J. R. Ciacci-Zanella
Labex-USA, EMBRAPA, Brazilian Agriculture Research Corporation, Brasilia, DF, Brazil

Current Topics in Microbiology and Immunology (2011) 370: 113–131
DOI: 10.1007/82_2011_196

Published Online: 22 January 2012

Contents

1 Brief Introduction to Influenza A Viruses

Influenza is a zoonotic viral disease that represents a health and economic threat to both human and animals worldwide. Influenza A viruses are the most studied of the *Orthomyxoviridae* since they can infect a large variety of birds and mammals including humans, pigs, horses, domestic poultry, marine mammals, cats, dogs and wild carnivores (Webster 2002; Thiry 2007). Wild aquatic birds were shown to be an asymptomatic reservoir for most subtypes of influenza A viruses (Scholtissek 1978; Fouchier et al. 2005). Moreover, influenza A virus ecology is intricate due to the high number of possible reassortment events and cross-species jumps that lead to their evolution (Webster et al. 1992). The hemagglutinin (HA) and the neuraminidase (NA) proteins encoded by gene segments 4 and 6, respectively, play a key role in the influenza life cycle and represent the primary targets of the host humoral immune response (Skehel and Wiley 2000). The HA protein is the most important determinant of virulence and host specificity as it binds to sialic acid-containing cell surface receptors on host epithelial cells (Shinya et al. 2006; Nicholls et al. 2008; Ayora-Talavera et al. 2009; de Wit et al. 2010). The HA mediates virus binding to *N*-acetylneuraminic acid-2,3-galactose (2,3-sialic acid) or *N*-acetylneuraminic acid-2,6-galactose (2, 6-sialic acid) terminal residues on sialyloligosaccharides for avian and mammalian virus primary binding predilection, respectively (Rogers and Paulson 1983). However, receptor binding restriction has been shown to be more complicated than previously understood, with tissues from human, swine and Japanese quail expressing both 2,3- and 2,6-sialic acid receptor types (Ito et al. 1998; Suzuki et al. 2000; Shinya et al. 2006; Wan and Perez 2006). Additionally, glycan array analysis has demonstrated that avian and mammalian adapted flu viruses can have binding spillover to the opposing receptor linkage type and that different strains bind preferentially to novel structures (such as sulphated and sialylated glycans) (Stevens et al. 2006). The NA is responsible for cleaving terminal sialic acid residues from carbohydrate moieties on the surfaces of the host cell and virus (Gottschalk 1957), thus assisting in virus cell entry by mucus degradation (Matrosovich et al. 2004) and the release and spread of progeny virions (Palese et al. 1974). The remaining six segments encode for the following structural and accessory proteins: PB2 (segment 1), PB1 (segment 2), PA (segment 3), NP (segment 5), M1 and M2 (segment 7), NS1 and NEP (segment 8) (Lamb and Krug 2007). Both HA and NA genes undergo two types of variation called antigenic drift and antigenic shift. Antigenic drift involves minor changes in the HA and NA due to polymerase errors during replication, whereas antigenic shift involves major changes in these molecules resulting from replacement of the entire gene segment as a consequence of reassortment events in the event that two (or more) unique viruses infect the same cell (Webster 1971). Based upon the major differences within the HA and NA proteins, 16 HA and 9 NA subtypes, naturally paired in different combinations, have been identified thus far (Webster et al. 1992; Rohm et al. 1996; Fouchier et al. 2005). Only a limited number of subtypes have been established in mammals. For example, only viruses of H1, H2, H3, N1 and N2 subtypes have circulated widely in the human population (Webster et al. 1992; Alexander and Brown 2000) and only H1, H3, N1 and N2 subtypes have been consistently isolated from pigs (Webster et al. 1992; Olsen 2002).

2 Influenza A Virus in Pigs

2.1 *First Detection in the United States*

Swine influenza was first recognized in pigs in the Midwestern U.S. in 1918 (Fig. 1) as a respiratory disease that coincided with the human pandemic known as the Spanish flu (Koen 1919). Since then, it has become an important disease to the swine industry throughout the world. The first influenza virus was isolated in 1930 by Shope (1931) and was demonstrated to cause respiratory disease in swine that was similar to human influenza. This strain was subsequently recognized as an H1N1 influenza virus, and swine were utilized in the following years as a model to study influenza pathogenesis in a natural host.

2.2 *Introduction of the Triple Reassortant H3N2 in Pigs*

Among the RNA viruses affecting mammals, influenza viruses and coronaviruses represent, as a consequence of different molecular mechanisms, two of the best examples of viruses with exceptionally plastic genomes. Thus, we should not be surprised that the high mutation and reassortment rates have propelled the evolution of influenza viruses in pigs in recent years. However, from the first characterization of swine influenza virus until the late 1990s, the classical swine lineage H1N1 (cH1N1) was relatively stable at the genetic and antigenic levels in

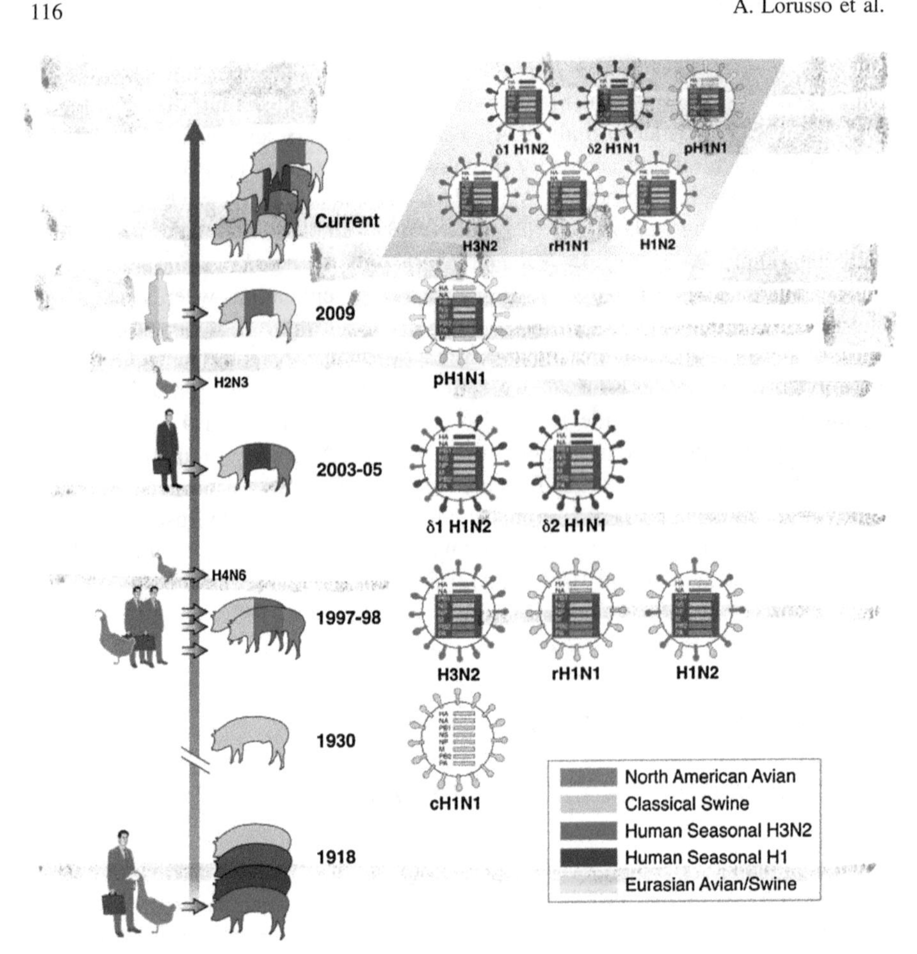

Fig. 1 Epidemiology and genetic composition of influenza viruses from U.S. and Canadian pigs. Swine virus lineage is color coded *pink*, avian lineage is coded *green*, human lineage is coded *blue* or *purple*. The chronology of transmission events leading to reassortant viruses with genes from swine, human and avian influenza virus lineages is visualized by the *vertical arrow*. The "Spanish flu" virus was transmitted from avian/human origin to pigs and evolved into the cH1N1, as indicated by the transition in color of pigs from *blue* to *light blue* to *red* to *pink*. The human and avian images to the left of the *vertical timeline* represent the species origin of viral gene segments donated to give rise to the swine influenza virus reassortants listed on the right side of the *vertical timeline*. *Time line* is not drawn to scale. Each viral subtype is represented with its eight gene segment arrangement. The triple reassortant H3N2 reassorted with the cH1N1 to produce rH1N1 and H1N2 subtypes with the triple reassortant internal gene (TRIG) cassette. Further reassortment events with two independent human H1 subtype viruses led to the $\delta 1$ H1N2 and $\delta 2$ H1N1. The source of the reassortment event producing the combination of gene segments in the 2009 pandemic H1N1 prior to its emergence in human and subsequent transmission from humans to pigs in 2009 is currently unknown. *Light green* indicates the Eurasian avian/swine lineage. The *gray* highlighted area illustrates the currently circulating influenza A subtypes in Canadian and American pigs

U.S. swine. Based on phylogenetic analysis, the cH1N1 lineage is closely related to the 1918 H1N1 Spanish flu virus (Easterday and van Reeth 1999) and other human influenza viruses isolated in the 1930s following the discovery of SIV. Although the cH1N1 was the predominant subtype causing disease in pigs until the late 1990s, there was serological evidence that human subtype H3 influenza viruses were circulating at a low frequency in U.S. pigs, but failed to establish a lineage with sustained transmission among swine (Chambers et al. 1991).

The epidemiology of influenza in pigs dramatically changed after the events of 1997–1998 (Fig. 1). In 1998, a severe influenza-like disease was observed in pigs in North Carolina with additional outbreaks in swine herds in Minnesota, Iowa and Texas. The causative agents for these outbreaks were identified as influenza A viruses of the H3N2 subtype. Genetic analysis of these H3N2 viruses showed that at least two different genotypes were present. The initial North Carolina isolate was a double reassortant and contained gene segments similar to those of the classical swine lineage (PB2, PA, NP, M, NS) combined with gene segments from a human seasonal H3N2 influenza virus circulating in 1995 (PB1, HA, NA). The isolates from Minnesota, Iowa and Texas were triple reassortants containing gene segments from the classical swine virus (NP, M, NS,) and the same human virus (PB1, HA, NA) in combination with an avian virus (PB2, PA) (Zhou et al. 1999). By the end of 1999, viruses antigenically and genetically related to the triple reassortant lineage were widespread in the U.S. swine population (Webby et al. 2000) whereas the double reassortant virus did not become established. Interestingly, the double and triple reassortant H3N2 viruses were shown to possess a similar HA encoding gene with identical residues in critical receptor binding regions, suggesting that their different successes were due to factors not associated with the HA and receptor binding pocket. The major difference between the two viruses was the acquisition of two avian polymerase genes (PB2 and PA) in the triple reassortant virus. The human lineage PB1, avian lineage PB2 and PA and swine lineage NP, M and NS found in contemporary swine influenza viruses are referred to as the triple reassortant internal gene (TRIG) constellation (Vincent et al. 2008). Genetic and antigenic evaluation of H3N2 swine influenza isolates since 1998 (Richt et al. 2003; Webby et al. 2004) indicate at least three introductions of human H3 subtype viruses became established in swine, leading to phylogenetic clusters I, II and III. The cluster III viruses have become dominant in North America (Gramer et al. 2007) and have continued to evolve into cluster III variants, also known as cluster IV (Olsen et al. 2006).

The H3N2 viruses not only evolved and became endemic in pigs but also reassorted with extant cH1N1 swine influenza viruses. The vast majority of the resulting reassortant and drift variant viruses since 1998 contain the TRIG. The H1N1 viruses containing the HA and NA from the cH1N1 virus and the TRIG from triple reassortant H3N2 viruses are referred as reassortant H1N1 (rH1N1) and the viruses containing the HA from the classical swine virus and the NA and TRIG from the triple reassortant H3N2 virus are H1N2 viruses (Karasin et al. 2002; Webby et al. 2004) (Fig. 1). Reassortant viruses have become endemic and co-circulate in most major swine producing regions of the U.S. and Canada, including further drift variants of H3N2 (Webby et al. 2000,2004; Richt et al. 2003; Olsen et al. 2006), H1N2 (Choi et al. 2002; Karasin et al. 2002), and rH1N1 (Webby et al. 2004). H3N1 viruses have occasionally been identified in limited outbreaks but do not appear to circulate widely (Lekcharoensuk et al. 2006; Ma et al. 2006). Moreover, the TRIG was shown to have accepted an avian lineage H2 and N3, producing a novel triple reassortant swine H2N3 in 2006 (Ma et al. 2007). More recently, introduction of H1 viruses with the HA gene of human H1N2 seasonal influenza virus origin (hu-like H1) that are genetically and antigenically distinct from the classical swine H1 lineage were reported in pigs in Canada (Karasin et al. 2006) (Fig. 1). Since 2005, hu-like H1N1 and H1N2 viruses containing the TRIG have emerged in swine herds across the U.S. (Vincent et al. 2009b) that have HA and NA segments most similar to H1N1 and H1N2 human seasonal influenza virus lineages from around 2003.

2.3 *Evolution of the H1 Subtype*

The well characterized contemporary swine influenza reassortant viruses possessing the ability to spread and become established in U.S. and Canadian swine populations have contained similar TRIG constellations. This would suggest that the TRIG constellation can accept multiple HA and NA types and may confer a selective advantage to viruses possessing this gene cassette (Bastien et al. 2010; Vijaykrishna et al. 2010). Moreover, since the acquisition of TRIG, an increase in the rate of mutation in North American swine influenza isolates appears to have occurred in H1 subtype hemagglutinins. Genetic mutation may be related to antigenic changes if mutations occur in antigenic sites of the HA, potentially resulting in escape from herd immunity. This scenario is in stark contrast with that observed with the cH1N1 viruses prior to acquiring TRIG. Indeed, cH1N1 viruses remained relatively stable genetically and antigenically for at least seven decades (Sheerar et al. 1989; Luoh et al. 1992; Noble et al. 1993; Olsen et al. 1993).

For best representing the evolution of the currently circulating H1 viruses, a cluster classification has been proposed (Fig. 2a). Viruses from the classical H1N1 lineage-HA acquired from the TRIG cassette evolved to form α-, β-, and γ-clusters based on the genetic makeup of the HA gene; whereas H1 subtypes strains with HA genes most similar to human seasonal H1 viruses form the δ-cluster (Vincent et al. 2009b). All four HA gene cluster types can be found with NA genes of either the N1 or N2 subtype. In order to study the evolution and the antigenic relationships among the H1 swine influenza virus subtypes, we recently analyzed 12 different strains, selected from the University of Minnesota Veterinary Diagnostic Laboratory (UMVDL) diagnostic case database (Lorusso et al. 2011). The viruses were isolated from outbreaks of respiratory disease in pigs from diagnostic cases submitted to the UMVDL in 2008 and are representative of each of the postulated four H1 clusters. All gene segments were sequenced and analyzed, and antigenic changes were measured for all twelve viruses using the hemagglutination inhibition (HI) assay and mapped by antigenic cartography. All 2008 H1 viruses contained the North American TRIG. Furthermore, variation was demonstrated in the six genes that make up the TRIG, but no HA cluster-specific patterns were detected among the genes composing the TRIG constellation. In contrast, an HA cluster-specific pattern was observed for the NA gene. The N1 gene of the α, β and γ cluster of the 2008 H1 viruses and of sequences publicly available each formed a separate clade within the North American N1 cluster. We speculate that the evolution of the H1 gene drives that of the N1 gene as well. Indeed, antigenic drifts that characterize the evolutionary history of the antigenic and phylogenetic clusters of H1 influenza virus in U.S. swine isolates were accompanied by changes in the N1 genes, thus allowing a parallel sub-cluster classification (Fig. 2b). A proper HA/NA pairing in association with the TRIG could optimize viral transmission and replication as shown by recent experiments in pigs. Indeed, experimental coinfection in the lower respiratory tract of inoculated pigs with two phylogenetically and genetically distant viruses, a triple reassortant H3N2 and cH1N1, resulted in the genesis, of all possible HA/NA combinations but only the parental H3N2 was found in two consecutive direct contact pig groups (Ma et al. 2010). These results confirm that multiple reassortments can occur but not all reassortants are readily transmissible.

The viruses representing the classical swine H1 lineage, phylogenetic clusters α, β and γ, had moderate to strong cross-reactivity within a cluster, especially within recent β- and γ-cluster viruses. However, cross-reactivity between clusters was more variable, ranging from no cross-reactivity to strong cross-reactivity, such as between α- and β-cluster viruses. This study suggested that the H1 is evolving by drift while maintaining the TRIG backbone, and that the resulting viruses differ genetically and antigenically with obvious consequences for vaccine and diagnostic test development. In 2008 and 2009, α cluster H1 viruses were rarely isolated from influenza outbreaks in pigs in North America, and while β cluster H1 viruses are still common, they occur with less frequency than the more dominant subtypes from the the γ and δ viruses. Since the acquisition of TRIG, the H1 of the classical swine lineage, under apparent evolutionary pressure, has developed multiple amino acid changes in the putative antigenic sites. The γ viruses are chronologically the newest H1 variants and it cannot be ruled out that the same mechanisms will be responsible for further H1 cluster variants. The genetic diversity within the H1 clusters was confirmed functionally by the demonstrated loss in cross-reactivity in the HI assay between H1 clusters overall. It is likely that, as a consequence of evolutionary and immunogical pressures, the H1 will continue to mutate in the future, allowing evasion of the immune system of the host or only partially protective immunity.

2.4 *Human-Like H1 Viruses*

Since 2005, H1N1 and H1N2 viruses with the HA gene derived from human viruses have spread across the U.S. in swine herds forming the δ-cluster H1 (Vincent et al. 2009b) (Fig. 1). The HAs from the human-like (hu) swine H1

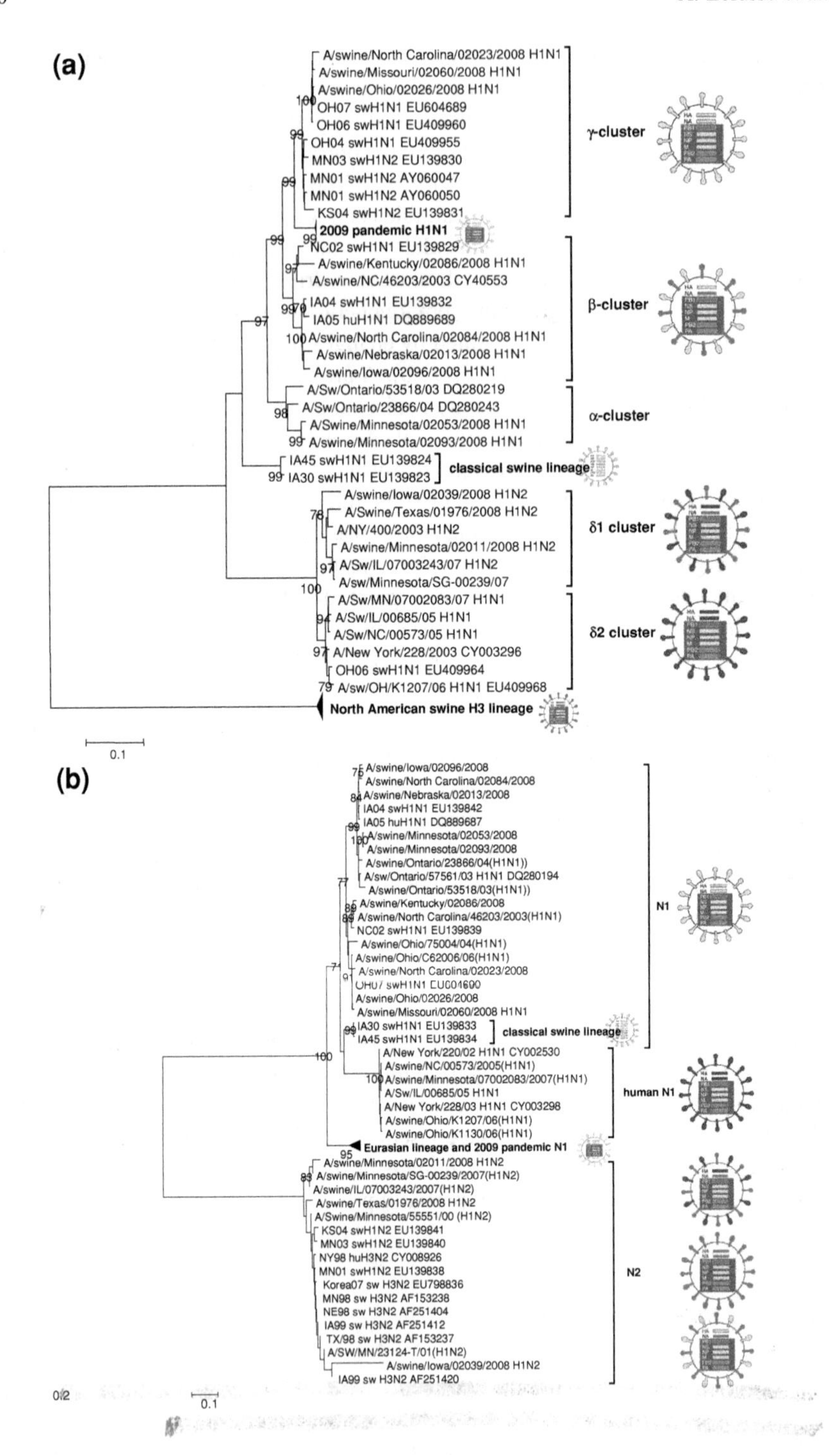
(a)
A/swine/North Carolina/02023/2008 H1N1
A/swine/Missouri/02060/2008 H1N1
A/swine/Ohio/02026/2008 H1N1
OH07 swH1N1 EU604689
OH06 swH1N1 EU409960
OH04 swH1N1 EU409955
MN03 swH1N2 EU139830
MN01 swH1N2 AY060047
MN01 swH1N2 AY060050
KS04 swH1N2 EU139831
2009 pandemic H1N1
γ-cluster
NC02 swH1N1 EU139829
A/swine/Kentucky/02086/2008 H1N1
A/swine/NC/46203/2003 CY40553
IA04 swH1N1 EU139832
IA05 huH1N1 DQ889689
A/swine/North Carolina/02084/2008 H1N1
A/swine/Nebraska/02013/2008 H1N1
A/swine/Iowa/02096/2008 H1N1
β-cluster
A/Sw/Ontario/53518/03 DQ280219
A/Sw/Ontario/23866/04 DQ280243
A/Swine/Minnesota/02053/2008 H1N1
A/swine/Minnesota/02093/2008 H1N1
α-cluster
IA45 swH1N1 EU139824
IA30 swH1N1 EU139823
classical swine lineage
A/swine/Iowa/02039/2008 H1N2
A/Swine/Texas/01976/2008 H1N2
A/NY/400/2003 H1N2
A/swine/Minnesota/02011/2008 H1N2
A/Sw/IL/07003243/07 H1N2
A/sw/Minnesota/SG-00239/07
δ1 cluster
A/Sw/MN/07002083/07 H1N1
A/Sw/IL/00685/05 H1N1
A/Sw/NC/00573/05 H1N1
A/New York/228/2003 CY003296
OH06 swH1N1 EU409964
A/sw/OH/K1207/06 H1N1 EU409968
δ2 cluster
North American swine H3 lineage
0.1
(b)
A/swine/Iowa/02096/2008
A/swine/North Carolina/02084/2008
A/swine/Nebraska/02013/2008
IA04 swH1N1 EU139842
IA05 huH1N1 DQ889687
A/swine/Minnesota/02053/2008
A/swine/Minnesota/02093/2008
A/swine/Ontario/23866/04(H1N1))
A/sw/Ontario/57561/03 H1N1 DQ280194
A/swine/Ontario/53518/03(H1N1))
A/swine/Kentucky/02086/2008
A/swine/North Carolina/46203/2003(H1N1)
NC02 swH1N1 EU139839
A/swine/Ohio/75004/04(H1N1)
A/swine/Ohio/C62006/06(H1N1)
A/swine/North Carolina/02023/2008
OH07 swH1N1 EU604690
A/swine/Ohio/02026/2008
A/swine/Missouri/02060/2008 H1N1
N1
IA30 swH1N1 EU139833
IA45 swH1N1 EU139834
classical swine lineage
A/New York/220/02 H1N1 CY002530
A/swine/NC/00573/2005(H1N1)
A/swine/Minnesota/07002083/2007(H1N1)
A/Sw/IL/00685/05 H1N1
A/New York/228/03 H1N1 CY003298
A/swine/Ohio/K1207/06(H1N1)
A/swine/Ohio/K1130/06(H1N1)
human N1
Eurasian lineage and 2009 pandemic N1
A/swine/Minnesota/02011/2008 H1N2
A/swine/Minnesota/SG-00239/2007(H1N2)
A/swine/IL/07003243/2007(H1N2)
A/swine/Texas/01976/2008 H1N2
A/Swine/Minnesota/55551/00 (H1N2)
KS04 swH1N2 EU139841
MN03 swH1N2 EU139840
NY98 huH3N2 CY008926
MN01 swH1N2 EU139838
Korea07 sw H3N2 EU798836
MN98 sw H3N2 AF153238
NE98 sw H3N2 AF251404
IA99 sw H3N2 AF251412
TX/98 sw H3N2 AF153237
A/SW/MN/23124-T/01(H1N2)
A/swine/Iowa/02039/2008 H1N2
IA99 sw H3N2 AF251420
N2
0.1

◂**Fig. 2** Neighbor-joining trees inferred from multiple nucleotide sequence alignment of segment 4 (HA, **a**) and segment 6 (NA, **2b**). **a** shows four H1 clusters of viruses, H1α, H1β, H1γ and H1δ (human-like H1) as indicated by the bars on the right of the tree. In both trees, the HA cluster specificity is indicated. The genomic constellation of each clade is indicated by the images on the right side of the tree. Classical swine lineage is color coded pink, avian lineage is coded green, human lineage is coded blue or purple. Light green indicates the Eurasian avian/swine lineage. Classical swine lineage-HA gene (**a**) was acquired by the TRIG cassette and evolved overtime to form α-, β- and γ- clusters. The introduction of human seasonal HA from H1N2 and H1N1 gave rise to δ cluster viruses differentiated phylogenetically by two distinct sub-clusters, δ1 and δ2 (**a**). Similar to the δ-cluster viruses in the HA phylogenetic analysis, β-viruses have split into two sub-clusters (**b**). Phylogenetic analyzes were conducted in MEGA4. Statistical support was provided by bootstrapping over 1,000 replicates and bootstrap values >70 are indicated at the correspondent node. The scale bars indicate the estimated numbers of nucleotide substitutions per site. human (*Hu*), swine (*Sw*)

viruses are genetically and antigenically distinct from classical swine lineage and derivatives. Indeed the putative antigenic site in the HA1 of the hu-like viruses possesses typical human lineage residues in contrast to that found in the HA1 of the α-, β- and γ-clusters (Lorusso et al. 2011). However, their TRIG genes are similar to those found in the TRIG cassette of the contemporary swine triple reassortant viruses (Vincent et al. 2009b). The HA from the δ-cluster viruses were shown to have most likely emerged from at least two separate introductions of human seasonal HA from H1N2 and H1N1 viruses being differentiated phylogenetically by two distinct sub-clusters, δ1 and δ2, respectively, (Lorusso et al. 2011; Vincent et al. 2009a). Viruses belonging to the δ-cluster were shown to be paired either with a N1 or N2 gene consistently of human lineage and not of swine lineage N1. δ1-subcluster viruses, first detected in 2003, showed an N2 gene preference whereas δ2-subcluster viruses, first detected in 2005, showed an N1 preference (Fig. 2b) initially but have subsequently begun to reassort. Limited HI cross-reactivity was demonstrated between the δ1 and δ2 viruses thus supporting the scenario assumed by the phylogenetic analysis (Fig. 2a). The hu-H1 viruses have become one of the major subtypes of influenza virus isolated and characterized from swine respiratory disease outbreaks. Indeed, if we consider the time period 2008–2010, the incidence of hu-H1 in swine respiratory disease outbreaks has dramatically increased. In 2008, 85% of the influenza viruses isolated from swine diagnostic cases submitted to the UMVDL were shown to be of the H1 subtype. Most of the H1 isolates (up to 78%) were of the γ- and β-cluster with the γ-cluster viruses found in slightly higher numbers, whereas δ-cluster viruses represented approximately 20% of the total. However, in 2009 the epidemiologic scenario changed. While the influenza A viruses isolated were mostly H1 subtype (five-fold more than the H3 subtype), the number of δ viruses now represented 40% of the total, thus quickly becoming the dominant subtype isolated from cases of respiratory disease. β- and γ-cluster viruses were 35 and 23% of the total H1 clusters represented, respectively. The same trend was shown in the early months of 2010 as well, with a slight increase in the number of δ-cluster viruses compared to the γ-cluster viruses, cluster IV H3 subtype viruses, and the newly emerged

2009 pandemic H1N1. An experimental in vivo study in 4-week-old pigs with an H1N1 isolate of the δ2-subcluster demonstrated differences in kinetics of lung lesion development, viral load in the lung and nasal shedding when compared to a virulent rH1N1 in the β-cluster. This study suggested the emerging virus genotype may not have been fully adapted to the swine host since virus replication in the lung and virus shedding from the nose were reduced compared to a contemporary rH1N1 (Vincent et al. 2009b). A more recent pathogenesis and transmission study in pigs comparing viruses in the δ1- and δ2-subclusters recapitulated the phenotypic differences seen in the initial study; however, the δ1-subcluster virus studied demonstrated increased virulence and nasal shedding over the δ2-subcluster viruses (Ciacci-Zanella, unpublished). Further studies are warranted in order to monitor the evolution of δ-cluster viruses. The presence of typical "human-like" residues in the receptor binding pocket in the HA of two of the δ-cluster viruses isolated in 2008 demonstrates that although these viruses have replicated in pigs for over five years, the swine viruses may preserve human-adapted receptor binding phenotypes (Lorusso et al. 2011). This preservation of human-like residues in the swine host may allow potential novel reassortant influenza viruses, including the δ-cluster swine viruses, to spill back into the human population. Escaping the immune response by changing the external makeup is a well-known strategy that influenza viruses adopt. The acquisition of human HA segments by the TRIG cassette platform were shown to be entirely different from those of the classical swine lineage and further drift derivatives provided an important antigenic advantage for these reassortant viruses. Indeed, the number of influenza outbreaks in which δ-cluster viruses were recognized as causative agents increased in the recent years. Moreover, geographical regions have differing cluster variants circulating, thus further complicating vaccine strain selection.

2.5 2009 Pandemic H1N1 in Pigs

In the early spring of 2009, the United States, Canada and Mexico reported community outbreaks of pneumonia in humans caused by a novel H1N1 influenza A virus. This virus subsequently spread across the globe at a high rate, prompting the WHO to declare a pandemic in June 2009 (Garten et al. 2009). Retrospectively, the earliest known case was identified February 24, 2009, in a baby from San Louis Potosi, Mexico (http://news.sciencemag.org/scienceinsider/2009/07/yet-another-new.html). This novel pandemic H1N1 possesses a unique genome with six gene segments (PB2, PB1, PA, HA, NP and NS) most closely related to the triple reassortant influenza viruses of the North American swine lineage, and the M and NA genes derived from a Eurasian lineage of swine influenza viruses (Dawood et al. 2009). The 2009 pandemic influenza became infamously known as "swine flu" due to the phylogenetic origin of the gene segments. However, since the recognition of the outbreak, infection in humans has not been connected to pig exposure (Dawood et al. 2009). Indeed, as it was believed to have occurred in 1918